Lecture Notes
in Control and I

Editors: M. Thoma, M. Morari

Frank J. Christophersen

Optimal Control of Constrained Piecewise Affine Systems

 Springer

Series Advisory Board

F. Allgöwer, P. Fleming, P. Kokotovic,
A.B. Kurzhanski, H. Kwakernaak,
A. Rantzer, J.N. Tsitsiklis

Author

Dr. Frank J. Christophersen
ETH Zurich
Automatic Control Laboratory
Physikstrasse 3
8092 Zurich
Switzerland
e-mail: fjc@control.ee.ethz.ch

Library of Congress Control Number: 2007927162

ISSN print edition: 0170-8643
ISSN electronic edition: 1610-7411
ISBN-10   3-540-72700-0 Springer Berlin Heidelberg New York
ISBN-13   978-3-540-72700-2 Springer Berlin Heidelberg New York

This work is subject to copyright. All rights are reserved, whether the whole or part of the material is concerned, specifically the rights of translation, reprinting, reuse of illustrations, recitation, broadcasting, reproduction on microfilm or in any other way, and storage in data banks. Duplication of this publication or parts thereof is permitted only under the provisions of the German Copyright Law of September 9, 1965, in its current version, and permission for use must always be obtained from Springer. Violations are liable for prosecution under the German Copyright Law.

Typesetting: by the authors and SPS using a Springer LATEX macro package

SPIN: 12027606   89/SPS   5 4 3 2 1 0

… for M. & J.

Noise is the key; life is full of noise. Only death is silent. ALEC EMPIRE

Preface

One of the most important and challenging problems in control is the derivation of systematic tools for the computation of controllers for general constrained nonlinear or *hybrid systems* [vS00] (combining continuous-valued dynamics with logic rules) that can guarantee (among others) closed-loop stability, feasibility, and optimality with respect to some objective function.

The focus of this book lies on the class of constrained discrete-time *piecewise affine* (PWA) *systems* [Son81]. This particular subclass of hybrid systems is obtained by partitioning the extended state-input space into regions and associating a different affine state update equation with each region. Under mild assumptions this modeling framework is equivalent to many other hybrid system classes reported in the literature [HDB01], and represents itself as a powerful modeling tool to capture general nonlinearities (e.g. by local approximation), constraints, saturations, switches, discrete inputs, and other hybrid modeling phenomena in dynamical systems.

Although piecewise affine systems are a subclass of general nonlinear systems, most of the (classical) control theory developed for nonlinear control does not apply, due to commonly made continuity and smoothness requirements.

The flexibility of this modeling framework and the recent technological advances in the fields of optimization and control theory have lead to a considerable interest in academia and industry in piecewise affine systems; not to mention that many engineering systems naturally express themselves or approximate nicely as piecewise affine systems, or can be 'straightforwardly' translated into them.

Simple examples are mechanical systems with backlash, where the system has a different dynamical behavior in the 'contact mode' than in the 'non-contact mode', or mechanical systems with dry friction (e.g. car tires), where the system is in the stick or slip mode, depending on the force acting on the systems. Another example is a chemical process that abruptly changes its phase, for instance, from the gaseous to the liquid phase, depending on the temperature or pressure prevailing in the system.

Naturally, the computation of stabilizing feedback controllers for constrained piecewise affine systems remains to be a challenging task, especially for fast sampled systems (i.e. in the range of milli- or microseconds).

The work at hand revolves around the efficient and systematic computation, analysis, and post-processing of *closed-form*, stabilizing, optimal, exact state-feedback controllers for these systems, where the cost function of the respective optimal control problem is composed of (piecewise) linear vector norms. For the here considered control problems, the underlying *constrained finite time optimal control* (CFTOC) problem can be solved by means of *multi-parametric (mixed-integer) linear programming* [Bor03, DP00]. Once the closed-form optimal solution is computed, the resulting controller can be utilized as optimal control lookup table of the current measured state. Thus the on-line computational effort reduces to a simple evaluation of the measured state in the lookup table to obtain the optimal control action.

In addition, this lookup table can be implemented for example into a microprocessor and cheaply be replicated in mass production. This is in contrast to the alternative on-line optimization counterpart, which usually implies expensive and large computational infrastructure or a limitation of the complexity of the corresponding optimization problem to cope with the possibly high sampling rates.

Outline

This book is structured as follows: Chapters 1 and 2 present some necessary background material on the mathematical terminology, the constrained (parametric) optimization, and systems and control theory, used throughout this manuscript.

The most successful modern control strategy both in theory and in practice for constrained systems is undoubtedly *Receding Horizon Control* (RHC) [MRRS00, Mac02], often also interchangeably called *Model (Based) Predictive Control* (MPC). In Chapter 3 the basic idea of receding horizon control is introduced, the underlying constrained finite time optimal control problem is explained and the corresponding closed-form solution and its usage is indicated.

Chapter 4 defines the discrete-time piecewise affine (PWA) system class that is the main focus of the whole work at hand.

The *constrained finite time optimal control* (CFTOC) problem with a (piecewise) linear vector norm based cost function for the class of PWA systems, and its use in the context of receding horizon control, is described in Chapter 6. One way to obtain the closed-form solution to this problem is by reformulating the problem into its equivalent mixed logical dynamical form and solving a multi-parametric mixed-integer linear program (mp-MILP) [Bor03, DP00]. This, however, is in general a computationally challenging task. Here a novel algorithm, which combines a dynamic programming strategy with a multi-parametric linear program solver and some basic polyhedral manipulation, is presented and compared to the aforementioned mp-MILP approach. By comparison with results in the literature it is shown that the presented algorithm solves the considered class of problems in a computationally more efficient manner.

Chapter 7 extends the ideas presented in Chapter 6 to the case of *constrained infinite time optimal control* (CITOC) problems for the same general system class. The equivalence of the dynamic programming generated solution with the solution to the infinite time optimal control problem is shown. Furthermore, new convergence results of the dynamic programming strategy for general nonlinear systems and stability guarantees for the resulting possibly *discontinuous* closed-loop system are given. A computationally efficient algorithm is presented, which obtains the closed-form infinite time optimal solution, based on a particular dynamic programming exploration strategy with a multi-parametric linear programming solver and basic polyhedral manipulation. Intermediate solutions of the dynamic programming strategy give stabilizing suboptimal controllers with guaranteed optimality bounds.

In Part III of this manuscript the further focus lies on the analysis and post-processing techniques for the closed-form optimal controllers obtained in Part II, i.e. Chapter 6 and 7.

Chapter 8 continues on the results proposed in [Pol95, KAS92] for computing linear vector norm based Lyapunov functions for linear discrete-time systems. The finitely terminating algorithm builds on a decomposition procedure and utilizes the solution of a finite and bounded sequence of feasibility LPs with very few constraints or, equivalently, very simple algebraic tests. The here computed weight W of the weighted vector norm is of small size and all the considered steps during the construction are elementary and scale well for large systems. Such Lyapunov functions are commonly utilized for a priori ensuring closed-loop asymptotic or exponential stability and all time feasibility when using a receding horizon control policy as considered in Chapter 3 and 6. For this purpose a simple algorithm for a class of PWA systems is presented.

In general, without extending or modifying the underlying optimization problem, stability and/or feasibility for all time of the closed-loop system is not guaranteed a priori, when a receding horizon control strategy is used. In Chapter 9 an algorithm is presented that by analyzing the CFTOC solution of a PWA system a posteriori extracts regions of the state space for which closed-loop stability and feasibility for all time can be guaranteed. The algorithm computes the maximal positively invariant set and stability region of a piecewise affine system by combining reachability analysis with some basic polyhedral manipulation. The simplicity of the overall computation stems from the fact that in all steps of the algorithm only linear programs need to be solved.

In Chapter 10 the concept of the stability tube is presented. By a posteriori analyzing a nominal stabilizing controller and the corresponding Lyapunov function of a general closed-loop nonlinear system one can compute a set, called stability tube, in the augmented state-input space. Any control chosen from this set guarantees closed-loop stability and a pre-specified level of closed-loop performance. Furthermore, these stability tubes can serve, for example, as a robustness analysis tool for the given nominal controller as well as a means to obtain an approximation of the nominal control law with lower complexity while preserving closed-loop stability and a certain performance level. For the PWA systems and PWA functions, considered in this manuscript, the overall computation of the stability tubes reduces to basic polytope manipulations.

The on-line evaluation of the above mentioned closed-form state feedback control law, i.e. the optimal lookup table, requires the determination of the state space region in which the measured state lies, in order to decide which 'piece' of the control law to apply. This procedure is called the *point location problem*, and the rate at which it can be solved determines the maximum sampling speed of the system allowed. In Chapter 11, a novel and computationally efficient algorithm based on bounding boxes and an n-dimensional interval tree is presented that significantly improves this point-location search for piecewise state feedback control laws defined over a large number of (possibly overlapping) polyhedra, where the required off-line preprocessing is low and so the approach can be applied to very large systems, which is substantiated by numerical examples.

Note on This Manuscript

This book is a revised and extended version of the author's PhD thesis [Chr06] written at the Automatic Control Laboratory of ETH Zurich in Switzerland. Parts of this manuscript are based on previously published articles. Note that most of the results in this manuscript were obtained in close collaboration with various colleagues. The appropriate cited references and list of authors throughout the text reflect this fact.

Acknowledgments

I would like to give a special thank you to Professor Manfred Morari for providing me with the most incredible infrastructure a researcher could ever wish for and his skill in bringing amazing, diverse, and interesting people from everywhere around the world together, which makes the working life at the Automatic Control Laboratory an extraordinary experience.

My deepest gratitude goes to my friend and mentor Mato Baotić who set me on the right track at the beginning of my research and who appreciated my pickiness. (BTW, I have full faith that one day you are going to meet a deadline yourself. ☺)

I'm indebted to a number of other people on the scientific level who have either collaborated with me, gave comments, or proofread this manuscript. Here they are: Colin N. Jones, Ulf Jönsson, Gianni Ferrari-Trecate, Johan Löfberg, Michal Kvasnica, Pascal Grieder, Miroslav Barić, Franta Kraus, Kristian Nolde, Thomas Besselmann, Andreas Kuhn, and Philipp Rostalski.

As mentioned before, the Automatic Control Laboratory is (and was) full of great people. In addition to the above mentioned ones, I want to thank Fabio D. Torrisi, Raphael Cagienard, Andrea G. Beccuti, Gültekin Erdem, Gabi Gentilini, Cristian Grossmann, Tobias Geyer, Marc Lawrence, Benji Hackl (you're the man; you rock!), Kari Unneland, John Lygeros, Pablo A. Parrilo, Dominik Niederberger, Urban Mäder, Antonello Caruso, Frauke Oldewurtel, George Papafotiou, Eugenio Cinquemani, Helfried Peyrl, Ioannis Lymperopoulos (a.k.a. the Master), Saša V. Raković, Valentina Sartori and Eleonora Zanderigo (not even here am I able to separate them by a simple comma), Stefan Richter, Ioannis Fotiou, Robert Riener, and not to forget my office mate Melanie N. Zeilinger. (Don't even try to figure out what this ordering means because there is none.)

Finally, I would like to thank (what I consider) my family Mutz, Jens, Lyly, Anıl, and Akın for simply everything.

Frank J. Christophersen
Zurich and Zagreb, 2006/'07

Contents

Preface VII
    Outline VIII
    Note on This Manuscript XI
    Acknowledgments XI

List of Definitions XVII

Notation XXI

I Background   1

1 Mathematical Necessities   3
    1.1   Set Terminology 3
    1.2   Polyhedra 7
        1.2.1   Basic Definitions 7
        1.2.2   Extensions 9
        1.2.3   Geometric Operations on Polyhedra 10
    1.3   Function Terminology 12

	1.4 Constrained Optimization	14
	1.5 Multi-Parametric Programming	17
2	SYSTEMS AND CONTROL THEORY	20
	2.1 Systems Formulation	20
	2.2 Invariance and Stability	21
3	RECEDING HORIZON CONTROL	26
	3.1 Basic Idea	27
	3.2 Closed-form Solution to Receding Horizon Control	28
	3.2.1 Time-varying Model and System	30
	3.2.2 Comments on the Closed-form Solution	31
	3.3 Choice of the Cost Function	32
	3.4 Stability and Feasibility in Receding Horizon Control	35
4	PIECEWISE AFFINE SYSTEMS	39
	4.1 Obtaining a PWA Model	41

II Optimal Control of Constrained Piecewise Affine Systems   43

5	INTRODUCTION	45
	5.1 Software Implementation	47
6	CONSTRAINED FINITE TIME OPTIMAL CONTROL	48
	6.1 CFTOC Solution via mp-MILP	51
	6.2 CFTOC Solution via Dynamic Programming	53
	6.3 An Efficient Algorithm for the CFTOC Solution	54
	6.4 Comments on the Dynamic Programming Approach for the CFTOC Problem	56
	6.5 Receding Horizon Control	58
	6.6 Examples	59
	6.6.1 Constrained PWA Sine-Cosine System	60
	6.6.2 Car on a PWA Hill	61
7	CONSTRAINED INFINITE TIME OPTIMAL CONTROL	64
	7.1 Problem Formulation	64
	7.2 Closed-loop Stability of the CITOC Solution	66
	7.3 CITOC Solution via Dynamic Programming	68
	7.3.1 Initial Value Function and Rate of Convergence	73
	7.3.2 Stabilizing Suboptimal Control	79
	7.4 An Efficient Algorithm for the CITOC Solution	80

	7.4.1 Alternative Choices of the Initial Value Function	84
7.5	Examples	86
	7.5.1 Constrained PWA Sine-Cosine System	86
	7.5.2 Constrained LTI System	88
	7.5.3 Car on a PWA Hill	89

III Analysis and Post-processing Techniques for Piecewise Affine Systems — 93

8 LINEAR VECTOR NORMS AS LYAPUNOV FUNCTIONS — 95
- 8.1 Introduction — 95
- 8.2 Software Implementation — 97
- 8.3 Stability Theory for Linear Systems Using Linear Vector Norms — 97
- 8.4 Complex W for Discrete-Time Systems — 98
- 8.5 Construction of a Real W for Discrete-Time Systems — 99
- 8.6 Simple Algebraic Test to Substitute the Feasibility LP (8.8) — 106
- 8.7 Computational Complexity and Finite Termination — 107
  - 8.7.1 Boundedness of m and Finite Termination — 107
  - 8.7.2 A Complexity Comparison — 109
- 8.8 Polyhedral Positive Invariant Sets — 110
- 8.9 Guaranteeing Exponential Stability in Receding Horizon Control for Piecewise Affine Systems — 110
- 8.10 Examples — 112

9 STABILITY ANALYSIS — 114
- 9.1 Introduction — 114
- 9.2 Constrained Finite Time Optimal Control of Piecewise Affine Systems — 115
- 9.3 Closed-Loop Stability and Feasibility — 116
  - 9.3.1 Computation of the Maximal Positively Invariant Set — 118
  - 9.3.2 Computation of the Lyapunov Stability Region — 123
- 9.4 Examples — 128
  - 9.4.1 Constrained Double Integrator — 128
  - 9.4.2 Constrained PWA Sine-Cosine System — 129
  - 9.4.3 Constrained PWA System — 131
  - 9.4.4 3-Dimensional Constrained PWA System — 133

10 STABILITY TUBES — 135
- 10.1 Introduction — 135
- 10.2 Constrained Finite Time Optimal Control of Piecewise Affine Systems — 136

10.3	Stability Tubes for Nonlinear Systems	139
10.4	Computation of Stability Tubes for Piecewise Affine Systems	143
10.5	Comments on Stability Tubes	144
10.6	Examples	147
	10.6.1 Constrained PWA Sine-Cosine System	147
	10.6.2 Car on a PWA Hill	148

11 Efficient Evaluation of Piecewise Control Laws Over a Large Number of Polyhedra   150

11.1	Introduction	150
11.2	Software Implementation	151
11.3	Point Location Problem	152
11.4	Constrained Finite Time Optimal Control of Piecewise Affine Systems	152
11.5	Alternative Search Approaches	154
11.6	The Proposed Search Algorithm	156
	11.6.1 Bounding Box Search Tree	157
	11.6.2 Local Search	159
11.7	Complexity	159
11.8	Examples	160
	11.8.1 Constrained LTI System	160
	11.8.2 Constrained PWA System	162
	11.8.3 Ball & Plate System	163

IV Appendix   167

A	Alternative Proof of Lemma 7.7(a)	169
	Bibliography	171
	Author Index	183
	Index	187

LIST OF DEFINITIONS

Definition 1.1 Convex set 3
Definition 1.2 Convex hull 3
Definition 1.3 Affine set 4
Definition 1.4 Affine hull 4
Definition 1.5 ε-Ball 4
Definition 1.6 Neighborhood 4
Definition 1.7 Dimension 4
Definition 1.8 Full-dimensional set 5
Definition 1.9 Affine dimension 5
Definition 1.10 Interior 5
Definition 1.11 Relative interior 5
Definition 1.12 Open set 5
Definition 1.13 Closed set 6
Definition 1.14 Closure 6
Definition 1.15 Boundary 6
Definition 1.16 Bounded set 6
Definition 1.17 Compact set 6
Definition 1.18 Redundancy 6
Definition 1.19 Set collection 7

Definition 1.20	Underlying set	7
Definition 1.21	Partition	7
Definition 1.22	Hyperplane	7
Definition 1.23	Half-space	8
Definition 1.24	Polyhedron	8
Definition 1.26	Face, facet, ridge, edge, vertex	8
Definition 1.27	Minimal representation	8
Definition 1.28	Polytope	8
Definition 1.31	Polytope collection	10
Definition 1.32	Polyhedral partition	10
Definition 1.33	Chebyshev ball	10
Definition 1.34	Projection	11
Definition 1.36	Affine function	12
Definition 1.37	Piecewise affine function	12
Definition 1.38	Polyhedral PWA function	12
Definition 1.39	Vector $1\text{-}/\infty$-norm	13
Definition 1.40	Matrix $1\text{-}/\infty$-norm	13
Definition 1.41	Convex/concave function	13
Definition 1.42	Positive/negative (semi-)definite function	13
Definition 1.43	Indefinite function	13
Definition 1.44	K-class function	14
Definition 1.45	$\widetilde{K}$-class function	14
Definition 1.46	K_∞-class function	14
Definition 1.47	L-class function	14
Definition 1.48	KL-class function	14
Definition 1.49	Convex optimization problem	15
Definition 1.51	Feasibility	18
Definition 2.1	Autonomous system	21
Definition 2.2	(Positively) Invariant set $\mathcal{O}$	21
Definition 2.3	Minimal/maximal positively invariant set	22
Definition 2.4	Input admissible set $\mathcal{X}^\mu$	22
Definition 2.5	(Maximal) control invariant set $\mathcal{C}_{\max}$	22
Definition 2.6	Attractivity	22
Definition 2.7	(Lyapunov) stability	23
Definition 2.8	Asymptotic stability	23
Definition 2.9	Exponential stability	23
Definition 2.10	Lyapunov function	23
Definition 2.12	Control Lyapunov function	24
Definition 3.4	Feasibility for all time	36
Definition 7.8	Operator $\mathbf{T}$ and $\mathbf{T}_\mu$	71
Definition 7.9	Stationarity	71

Definition 8.3   Eigenvalue multiplicity 98
Definition 9.3   Maximal positively invariant set $\mathcal{O}_{\max}$ 117
Definition 9.4   Region of attraction $\mathcal{A}$ 118
Definition 9.5   Lyapunov stability region $\mathcal{V}$ 118
Definition 10.3 Stability tube 140

Notation

Throughout this book, as a general rule, scalars and vectors are denoted with lower case letters ($a, b, \ldots, \alpha, \beta, \ldots$), matrices are denoted with upper case letters ($A, B, \ldots$), and sets are denoted with upper case calligraphic letters ($\mathcal{A}, \mathcal{B}, \ldots$).

In the following let $i, j, m, n \in \mathbb{N}$ be integers such that $i \leq m, j \leq n$, let $s \in \mathbb{R}^n$ be a column vector, $S \in \mathbb{R}^{n \times n}$ be a square matrix, $R \in \mathbb{R}^{m \times n}$ be a rectangular matrix, $\mathcal{I} \subseteq \{1, \ldots, m\}$ be a set of integers, and $\mathcal{S} \subseteq \mathbb{R}^n$ be a subset or collection of subsets of an n-dimensional real vector space.

General Operators, Relations, and Functions

$\cdot$	general placeholder (for any variable, index, set, or space)
$\tilde{\cdot}$	function, variable, set, etc. that was derived or is in a special relationship to $\cdot$
$\check{\cdot}$	generic object (set, function, etc.)
$\ldots$	"and so forth"
$:=$	left-hand side is defined by the right-hand side
$=:$	right-hand side is defined by the left-hand side
$\mid$	such that
$:$	such that

$\in$	is element of (belongs to)
$\forall$	for all
$\exists$	there exists at least one
$\exists!$	there exists exactly one
$\notin, \nexists, \ldots$	/ denotes negation
$\cdot \to \cdot$	mapping, "maps from ... to ..."
$\{\cdot, \cdot, \ldots\}$	a set or sequence
$(\cdot, \cdot, \ldots)$	a composite variable (ordered list, tuple), e.g. $(s, i, \mathcal{S}) \in \mathbb{R}^n \times \mathbb{N} \times 2^{\mathbb{R}^m}$
$\vee$	or
$\wedge$	and
$\Rightarrow$	implies
$\Leftarrow$	is implied by
$\Leftrightarrow$	equivalence, "if and only if"
$\mathring{f}(\cdot)$	restriction of the function $f(\cdot)$ to the neighborhood around the origin
$\deg(\cdot)$	degree of a polynomial
$\dom(\cdot)$	domain of a function/mapping
$\range(\cdot)$	range of a function/mapping
$\lceil \cdot \rceil$	ceiling function, i.e. smallest integer greater or equal than its argument
$\lfloor \cdot \rfloor$	floor function, i.e. greatest integer less or equal than its argument
j	$\sqrt{-1}$
e	Euler number, $e \approx 2.71828182845905$

Sets and Spaces

$\{\cdot, \ldots\}$	set or sequence
$[\cdot, \cdot)$	interval on the real line, where here the lower endpoint is included (closed) and the upper endpoint in not (open)
$\mathcal{B}_\varepsilon(x)$	n_x-dimensional ε-ball around $x \in \mathbb{R}^{n_x}$
$\mathcal{N}(\mathcal{S})$	neighborhood of the set $\mathcal{S}$
$\emptyset$	empty-set
$\mathbb{C}$	complex numbers
$\mathbb{C}^n$	space of n-dimensional (column) vectors with complex entries
$\mathbb{C}^{m \times n}$	space of m by n matrices with complex entries

Notation

$\mathbb{R}$	real numbers
$\mathbb{R}^n$	space of n-dimensional (column) vectors with real entries
$\mathbb{R}^{m \times n}$	space of m by n matrices with real entries
$\mathbb{R}_{\geq c}$	real numbers $\geq c$, i.e. $\mathbb{R}_{\geq c} := \{x \in \mathbb{R} \mid x \geq c\}$
$\mathbb{Q}$	rational numbers
$\mathbb{Z}$	integers
$\mathbb{Z}_{\geq c}$	integers $\geq c$, i.e $\mathbb{Z}_{\geq c} := \{x \in \mathbb{Z} \mid x \geq c\}$
$\mathbb{N}$	natural numbers (positive integers), $\mathbb{N} = \mathbb{Z}_{>0}$
$\mathbb{N}_{\geq c}$	natural numbers $\geq c$, i.e $\mathbb{N}_{\geq c} := \{x \in \mathbb{N} \mid x \geq c\}$

Set Operators

$(\subset) \subseteq$	(strict) subset
$(\supset) \supseteq$	(strict) superset
$\cap$	intersection
$\cup$	union
$\setminus$	set difference
$\times$	Cartesian product, $\mathcal{X} \times \mathcal{Y} = \{(x,y) \mid x \in \mathcal{X}, y \in \mathcal{Y}\}$
$\lvert \mathcal{I} \rvert$	number of elements (cardinality) of a set $\mathcal{I}$
$N_{\mathcal{S}}$	number of polytopes (or sets) in the collection of polytopes (resp. sets), $\mathcal{S} = \{\mathcal{S}_i\}_{i=1}^{N_{\mathcal{S}}}$
$2^{\mathcal{S}}$	power set (set of all subsets of $\mathcal{S}$)
$\mathcal{S}_i$	i-th polytope (or set) of a collection of polytopes (resp. sets) $\mathcal{S} = \{\mathcal{S}_i\}_{i=1}^{N_{\mathcal{S}}}$
$\mathring{\mathcal{S}}$	collection of sets $\mathcal{S}_i$ for which $0 \in \mathcal{S}_i$ (neighborhood set around the origin)
$\partial \mathcal{S}$	boundary of $\mathcal{S}$
$\bar{\mathcal{S}}$	closure of $\mathcal{S}$
$\underline{\mathcal{S}}$	underlying set of $\mathcal{S}$
$\mathrm{conv}(\cdot)$	convex hull
$\mathrm{aff}(\cdot)$	affine hull
$\dim(\cdot)$	dimension
$\mathrm{affdim}(\cdot)$	affine dimension
$\mathrm{int}(\cdot)$	strict interior
$\mathrm{relint}(\cdot)$	relative interior
$\mathrm{vert}(\cdot)$	set of vertices
$\mathrm{vol}(\cdot)$	volume
$\mathrm{face}_i(\mathcal{S})$	i-th $(n-1)$-dimens. face of the n-dimensional polytope $\mathcal{S}$
$\mathrm{face}_i^d(\mathcal{S})$	i-th d-dimensional face of the n-dimensional polytope $\mathcal{S}$

$\text{proj}_{\mathcal{P}}(\cdot)$	projection onto the constraint or set $\mathcal{P}$
$\text{proj}_{x}(\cdot)$	projection onto the x-subspace

Operators on Vectors and Matrices

$\mathbb{0}_n$	vector of zeros, $\mathbb{0}_n := [0\,0\,\ldots\,0]' \in \mathbb{R}^n$
$\mathbb{1}_n$	vector of ones, $\mathbb{1}_n := [1\,1\,\ldots\,1]' \in \mathbb{R}^n$
e_i, e_i^n	i-th column of the unit-matrix I (resp. I_n)
$[\cdot, \cdot, \ldots]$	a matrix (or a vector)
$<, \leq, =, \geq, >$	element-wise comparison of vectors
$\text{diag}(s)$	diagonal matrix with diagonal elements $[\text{diag}(s)]_{i,i} = s_i$
s'	row vector, transpose of a vector
$s_i, [s]_i$	i-th element of a vector (second notation only if ambiguous)
$s_{\mathcal{I}}, [s]_{\mathcal{I}}$	vector formed from the elements indexed by $\mathcal{I}$
$|s|$	element-wise absolute value
$\|s\|$	(any) vector norm
$\|s\|_p$	vector p-norm (Hölder norm)
$\|s\|_1$	vector 1-norm (sum of absolute elements of a vector, Manhattan norm)
$\|s\|_2$	vector 2-norm (Euclidian norm)
$\|s\|_\infty$	vector ∞-norm (largest absolute element of a vector)
0	zero matrix (of appropriate dimension)
$0_{m \times n}$	zero matrix of dimension $m \times n$
I	identity matrix (of appropriate dimension)
I_n	identity matrix of dimension $n \times n$
$[\cdot, \cdot, \ldots]$	a matrix (or a vector)
$R_{i,j}, [R]_{i,j}$	(i,j)-th element of a matrix (second notation only if ambiguous)
$[R]_i, [R]_{i,\cdot}$	i-th row of a matrix
$[R]_{\mathcal{I}}$	matrix formed from the rows indexed by $\mathcal{I}$
$[R]_{\cdot,j}$	j-th column of a matrix
$[R]_{\cdot,\mathcal{J}}$	matrix formed from the columns indexed by $\mathcal{J}$
$S\,(\succeq)\,\succ 0$	positive (semi)definite matrix
$S\,(\preceq)\,\prec 0$	negative (semi)definite matrix
R'	transpose of a matrix
R^H	conjugate transpose (Hermitian adjoint) of a matrix
S^{-1}	inverse of the square matrix
$\det(S)$	determinant of the square matrix
$\text{diag}(S)$	vector composed of the diagonal elements of a matrix, $[\text{diag}(S)]_i = S_{i,i}$

$\mathrm{diag}(R_1, \ldots)$	block diagonal matrix composed of the matrices R_i
$\mathrm{rank}(R)$	rank of a matrix
$\mathrm{range}(R)$	the subspace spanned by the columns of R
$\mathrm{kern}(R)$	the (right) null space of R, $\{s \in \mathbb{R}^n \mid Rs = 0\}$
$\lambda(R)$	eigenvalue of a matrix, vector of eigenvalues of a matrix
$\rho(R)$	spectral radius of a matrix
$\|R\|$	(any) matrix norm
$\|R\|_p$	induced matrix p-norm
$\|R\|_1$	induced matrix 1-norm (maximum column-sum norm)
$\|R\|_2$	induced matrix 2-norm (max. singular value, spectral norm)
$\|R\|_\infty$	induced matrix ∞-norm (maximum row-sum norm)

Optimization and Complexity

inf	infimum
max	maximum
min	minimum
sup	supremum
φ^*	at optimum of φ
O	order of complexity (standard Landau notation [HW79])

Systems and Control Theory

n_x	number of states, $n_x \in \mathbb{N}$
n_u	number of inputs, $n_u \in \mathbb{N}$
n_w	number of additive disturbance inputs, $n_w \in \mathbb{N}$
$N_\mathcal{D}$	number of system dynamics, $N_\mathcal{D} \in \mathbb{N}$
i, j	index of a set or region in the state or input space
d	index of a dynamic in the extended state-input space
k	iteration index of a dynamic program
x	state vector, $x \in \mathbb{R}^{n_x}$
x^+	successor state vector, i.e. if $x := x(t)$ then $x^+ := x(t+1)$
$x(t+\tau\|t)$	predicted state vector at time $t + \tau$ based on the 'real' state vector $x(t)$ at time t and a given system prediction model
x_t	measured state at time instance t
u	input/control vector (also called 'control action'), $u \in \mathbb{R}^{n_u}$
w	additive disturbance vector, $w \in \mathbb{R}^{n_w}$
t	time, $t \in \mathbb{Z}$

T	prediction horizon, $T \in \mathbb{N}$
U_T	sequence of input vectors, $U_T := \{u(t)\}_{t=0}^{T-1}$
$U_T^*(x(0))$	opt. sequence of input vectors based on the initial state $x(0)$
f	state update function, $f : \mathbb{X} \times \mathbb{U} \times \mathbb{T} \to \mathbb{X}$
f^{CL}	closed-loop state update function, $f^{\text{CL}} : \mathbb{X} \times \mathbb{T} \to \mathbb{X}$
μ	state feedback control law, $\mu : \mathbb{X} \times \mathbb{T} \to \mathbb{U}$
ℓ	stage cost function, $\ell : \mathbb{X} \times \mathbb{U} \times \mathbb{T} \to \mathbb{R}$
ℓ_T	terminal cost function (also called final penalty cost function), $\ell_T : \mathbb{X} \to \mathbb{R}$
J	cost function or objective function, $J : \mathbb{X} \times \mathbb{U} \to \mathbb{R}$
J^*	value function, i.e. the optimal cost function, $J^* : \mathbb{X} \to \mathbb{R}$
V	Lyapunov function, $V : \mathbb{X} \to \mathbb{R}$
$\mathbf{T}$	dynamic programming operator, see Definition 7.8
$\mathbb{X}$	pre-specified set of state vectors, $\mathbb{X} \subseteq \mathbb{R}^{n_x}$
$\mathbb{U}$	pre-specified set of input vectors, $\mathbb{U} \subseteq \mathbb{R}^{n_u}$
$\mathbb{D}$	pre-specified set of state-input vectors, $\mathbb{D} \subseteq \mathbb{R}^{n_x+n_u}$
$\mathbb{T}$	pre-specified time (interval), $\mathbb{T} \subseteq \mathbb{Z}$
$\mathcal{X}$	(arbitrary) set of state vectors, $\mathcal{X} \subseteq \mathbb{R}^{n_x}$
$\mathcal{X}^f$	terminal target set of state vectors, i.e. $x(T) \in \mathcal{X}^f \subseteq \mathbb{R}^{n_x}$
$\mathcal{X}_T$	set of feasible state vectors for a CFTOC problem with prediction horizon T, $\mathcal{X}_T \subseteq \mathbb{R}^{n_x}$
$\mathcal{U}$	(arbitrary) set of input vectors, $\mathcal{U} \subseteq \mathbb{R}^{n_u}$
$\mathcal{W}$	(arbitrary) set of additive disturbance vectors, $\mathcal{W} \subseteq \mathbb{R}^{n_w}$
$\mathcal{D}$	set of state-input vectors, state-update function domain $\mathcal{D} \subseteq \mathbb{R}^{n_x+n_u}$
$\mathcal{V}$	stability region in the Lyapunov sense for an autonomous system, $\mathcal{V} \subseteq \mathbb{R}^{n_x}$
$\mathcal{A}$	region of attraction of an autonomous system, $\mathcal{A} \subseteq \mathbb{R}^{n_x}$
$\mathcal{O}$	(positively) invariant set for an auton. system, $\mathcal{O} \subseteq \mathbb{R}^{n_x}$
$\mathcal{O}_{\min}$	minimal positively invariant set for an autonomous system, $\mathcal{O}_{\min} \subseteq \mathcal{O} \subseteq \mathbb{R}^{n_x}$
$\mathcal{O}_{\max}$	maximal positively invariant set for an autonomous system, $\mathcal{O}_{\max} \subseteq \mathbb{R}^{n_x}$
$\mathcal{X}^\mu$	input admissible set, $\mathcal{X}^\mu \subseteq \mathbb{R}^{n_x}$

$\mathcal{S}(V,\beta)$ stability tube, set of stabilizing state-input vectors with respect to Lyapunov function $V(\cdot)$ and Lyapunov decay bound $\beta(\cdot)$, $\mathcal{S}(V,\beta) \subseteq \mathbb{R}^{n_x+n_u}$

Acronyms and Indices

CFTOC	constrained finite time optimal control
CITOC	constrained infinite time optimal control
CL	closed-loop
CLF	control Lyapunov function
DP	dynamic program(ming)
JF	Jordan form
LMI	linear matrix inequality
LP	linear program(ming)
LQR	linear quadratic regulator
LTI	linear time invariant
MILP	mixed integer linear program
MPC	model predictive control
mp-LP	multi-parametric linear program
mp-QP	multi-parametric quadratic program
mp-MILP	multi-parametric mixed integer linear program
OL	open-loop
PWA	piecewise affine
PWQ	piecewise quadratic
QP	quadratic program(ming)
RH	receding horizon
RHC	receding horizon control
RJF	real Jordan form
SDP	semi definite program(ming)

PART I

BACKGROUND

1
MATHEMATICAL NECESSITIES

For the sake of completeness, reference, and full understanding of the work described in the later chapters, standard (as well as some non-standard) mathematical concepts are given in this chapter. Most of the definitions are well established and can be found in mathematical textbooks on analysis [Roc97], metric theory [BBI01], optimization [Ber95, BS79], and stability theory [Hah67, Vid93, Kha96]. Other definitions are slightly adopted for the framework of this manuscript.

1.1 Set Terminology

Definition 1.1 (Convex set). A set $S \subseteq \mathbb{R}^{n_s}$ is *convex* if the line segment connecting any pair of points of S lies entirely in S, i.e.

$$s_1, s_2 \in S, \quad 0 \leq \alpha \leq 1 \quad \Rightarrow \quad \alpha s_1 + (1-\alpha) s_2 \in S.$$

Definition 1.2 (Convex hull). The *convex hull* of a set $S \subseteq \mathbb{R}^{n_s}$ is the smallest convex set containing S, i.e.

$$\mathrm{conv}(S) := \bigcap \left\{ \tilde{S} \subseteq \mathbb{R}^{n_s} \mid S \subseteq \tilde{S}, \, \tilde{S} \text{ is convex} \right\}.$$

For any finite set $\mathcal{S} = \{s_1, \ldots, s_k\}$ with $k \in \mathbb{N}$ it follows that

$$\text{conv}(\mathcal{S}) = \left\{ \sum_{i=1}^{k} \alpha_i s_i \;\middle|\; \forall s_i \in \mathcal{S},\, \alpha_i \geq 0,\, i = 1, \ldots, k,\, \sum_{i=1}^{k} \alpha_i = 1,\, k \in \mathbb{N} \right\}.$$

Definition 1.3 (Affine set). A set $\mathcal{S} \subseteq \mathbb{R}^{n_s}$ is *affine* if the line through any two points in $\mathcal{S}$ lies in $\mathcal{S}$, i.e.

$$s_1, s_2 \in \mathcal{S},\quad s_1 \neq s_2,\quad \alpha \in \mathbb{R} \quad \Rightarrow \quad (1 - \alpha)s_1 + \alpha s_2 \in \mathcal{S}. \qquad \square$$

Definition 1.4 (Affine hull). The *affine hull* of a set $\mathcal{S} \subseteq \mathbb{R}^{n_s}$ is the smallest affine set containing $\mathcal{S}$, i.e.

$$\text{aff}(\mathcal{S}) := \bigcap \left\{ \tilde{\mathcal{S}} \subseteq \mathbb{R}^{n_s} \;\middle|\; \mathcal{S} \subseteq \tilde{\mathcal{S}},\, \tilde{\mathcal{S}} \text{ is affine} \right\}. \qquad \square$$

For any finite set $\mathcal{S} = \{s_1, \ldots, s_k\}$ with $k \in \mathbb{N}$ it follows that

$$\text{aff}(\mathcal{S}) = \left\{ \sum_{i=1}^{k} \alpha_i s_i \;\middle|\; \forall s_i \in \mathcal{S},\, \alpha_i \in \mathbb{R},\, i = 1, \ldots, k,\, \sum_{i=1}^{k} \alpha_i = 1,\, k \in \mathbb{N} \right\}.$$

Definition 1.5 (ε-Ball). The open n_x-dimensional ε-ball in $\mathbb{R}^{n_x}$ around a given (center) point $x_c \in \mathbb{R}^{n_x}$ is the set

$$\mathcal{B}_\varepsilon(x_c) := \{x \in \mathbb{R}^{n_x} \mid \|x - x_c\| < \varepsilon\},$$

where the radius $\varepsilon > 0$ and $\|\cdot\|$ denotes any vector norm (usually the Euclidean vector norm $\|\cdot\|_2$). $\qquad \square$

A more general concept can be given as follows

Definition 1.6 (Neighborhood). The *neighborhood* of a subset $\mathcal{S}$ of $\mathcal{X} \subseteq \mathbb{R}^{n_x}$ is defined as a set $\mathcal{N}(\mathcal{S})$ with $\mathcal{S} \subset \mathcal{N}(\mathcal{S}) \subseteq \mathcal{X}$ such that for each $s \in \mathcal{S}$ there exists an n_x-dimensional ε-ball with $\mathcal{B}_\varepsilon(s) \subseteq \mathcal{N}(\mathcal{S})$ and $\varepsilon > 0$. $\qquad \square$

Clearly, the neighborhood of a point $x_c \in \mathcal{X} \subseteq \mathbb{R}^{n_x}$ is a set $\mathcal{N}(x_c) \subseteq \mathcal{X}$ such that there exists an n_x-dimensional ε-ball with $\mathcal{B}_\varepsilon(x_c) \subseteq \mathcal{N}(x_c)$ and $\varepsilon > 0$.

Definition 1.7 (Dimension). Let $n(s) \in \mathbb{N}_{\geq 0}$ be a finite integer for $s \in \mathcal{S} \subseteq \mathbb{R}^{n_s}$, where $n_\mathcal{S} \in \mathbb{N}_{\geq 0}$ is finite. The *local dimension* $n(s)$ of $\mathcal{S}$ at the point s is the largest number $n(s)$ of an $n(s)$-dimensional ε-ball $\mathcal{B}_\varepsilon(\tilde{s})$ around $\tilde{s}$ with $\mathcal{B}_\varepsilon(\tilde{s}) \subseteq \tilde{\mathcal{S}}$ for $\varepsilon \to 0$, where $\tilde{\mathcal{S}}$ is a Euclidean space which is locally isomorphic to $\mathcal{S}$ around s, and $\tilde{s} \in \tilde{\mathcal{S}}$ is the corresponding map of s.

The *dimension* of $\mathcal{S}$ is defined as

$$\dim(\mathcal{S}) := \sup_{s \in \mathcal{S}} n(s),$$

and the dimension of the empty-set is by convention $\dim(\emptyset) := -1$. $\qquad \square$

Definition 1.7 for dimension is often also referred to as manifold's or geometric dimension and restricts the general notion of dimension to normed vector spaces, which is sufficient for the work at hand. In the case that S is a smooth manifold, $\tilde{S}$ can be viewed as the local tangent-space to S in the point s. For example, the dimension of a circle in $\mathbb{R}^2$ is one, the dimension of an affine plane in $\mathbb{R}^3$ is two, any countable, non-empty set has dimension zero, and the dimension of two parallel lines in $\mathbb{R}^2$ is one.

Definition 1.8 (Full-dimensional set). A set $S \subseteq \mathbb{R}^{n_s}$ is *full-dimensional* if $\dim(S) = n_s$, i.e.

$$\exists \varepsilon > 0, s \in \mathbb{R}^{n_s} \quad \text{such that} \quad \mathcal{B}_\varepsilon(s) \subseteq S.$$

If $\dim(S) < n_s$, then S is called *lower-dimensional*. □

Thus, the just above mentioned examples are not full-dimensional sets but lower-dimensional.

Often it is useful to embed a set into a higher dimensional space. For this purpose let us define

Definition 1.9 (Affine dimension). The affine dimension of a set $S \subseteq \mathbb{R}^{n_s}$ is the dimension of its affine hull, i.e.

$$\text{affdim}(S) := \dim(\text{aff}(S)). $$

□

Therefore, the circle or the parallel lines in $\mathbb{R}^2$ has an affine dimension of two.

Definition 1.10 (Interior). The *interior* of a set $S \subseteq \mathbb{R}^{n_s}$ is the set of all points for which there exist an n_s-dimensional ε-ball inscribed in S, i.e.

$$\text{int}(S) := \{s \in S \mid \exists \varepsilon > 0, \, \mathcal{B}_\varepsilon(s) \subseteq S\}.$$

□

According to the above definition, any non full-dimensional set in $\mathbb{R}^{n_s}$ has an empty interior. For such sets it is useful to define the concept of *relative interior*.

Definition 1.11 (Relative interior). The *relative interior* of a set $S \subseteq \mathbb{R}^{n_s}$ is defined by

$$\text{relint}(S) := \{s \in \text{aff}(S) \mid \exists \varepsilon > 0, \, \mathcal{B}_\varepsilon(s) \cap \text{aff}(S) \subseteq S\}.$$

□

Definition 1.12 (Open set). A set $S \subseteq \mathbb{R}^{n_s}$ is *open* if it is equal to its interior, i.e.

$$S = \text{int}(S).$$

□

Note that for a point $x \in \mathbb{R}^{n_x}$ the ball $\mathcal{B}_\varepsilon(x)$ is, by definition, an open, non-empty, full-dimensional set in $\mathbb{R}^{n_x}$.

Definition 1.13 (Closed set). A set $\mathcal{S} \subseteq \mathbb{R}^{n_s}$ is *closed* if every point not in $\mathcal{S}$ has a neighborhood disjoint from $\mathcal{S}$, i.e.

$$\forall x \notin \mathcal{S}, \ \exists \varepsilon > 0 \quad \text{such that} \quad \mathcal{B}_\varepsilon(x) \cap \mathcal{S} = \emptyset. \qquad \square$$

The definitions of open and closed sets are not mutually exclusive, e.g. sets such as $\{x \in \mathbb{R} \mid a < x \leq b\}$ are neither closed nor open and are usually referred to as *half-closed* or *half-open* sets. Other sets like the empty-set $\emptyset$ or the real vector space $\mathbb{R}^n$, depending on the topological space chosen, are considered to be both open and closed. The *extended real vector space* is usually denoted by $\overline{\mathbb{R}} := \mathbb{R} \cup \{+\infty\} \cup \{-\infty\}$.

Definition 1.14 (Closure). The *closure* of a set $\mathcal{S} \subseteq \mathbb{R}^{n_x}$ is defined as a unique smallest closed set $\bar{\mathcal{S}}$ containing $\mathcal{S}$ or, equivalently,

$$\bar{\mathcal{S}} := \{s \in \mathbb{R}^{n_s} \mid \forall \varepsilon > 0, \ \mathcal{B}_\varepsilon(s) \cap \mathcal{S} \neq \emptyset\}. \qquad \square$$

It follows immediately that

$$\text{relint}(\mathcal{S}) \subseteq \mathcal{S} \subseteq \bar{\mathcal{S}}.$$

Definition 1.15 (Boundary). The *boundary* of a set $\mathcal{S} \subseteq \mathbb{R}^{n_s}$ is defined as

$$\partial \mathcal{S} := \bar{\mathcal{S}} \setminus \text{int}(\mathcal{S}). \qquad \square$$

Definition 1.16 (Bounded set). A set $\mathcal{S} \subseteq \mathbb{R}^{n_s}$ is *bounded* if it is contained inside some ball $\mathcal{B}_r(\cdot)$ of finite radius r, i.e.

$$\exists r < \infty, \ s \in \mathbb{R}^{n_s} \quad \text{such that} \quad \mathcal{S} \subseteq \mathcal{B}_r(s). \qquad \square$$

Definition 1.17 (Compact set). A set $\mathcal{S}$ is *compact* if it is closed and bounded. $\square$

Consequently it follows that for any set $\mathcal{S} \subseteq \mathbb{R}^{n_s}$ the following holds

$$\text{int}(\mathcal{S}) \subseteq \text{relint}(\mathcal{S}) \subseteq \mathcal{S} \subseteq \text{conv}(\mathcal{S}) \subseteq \text{aff}(\mathcal{S}) \quad \text{and} \quad \mathcal{S} \subseteq \bar{\mathcal{S}} \subseteq \text{aff}(\mathcal{S}).$$

Definition 1.18 (Redundancy). Let the set $\mathcal{S} \subseteq \mathbb{R}^{n_s}$ be defined by the constraints $s_i(x) \leq 0$ with $i = 1, \ldots, N_s$, i.e.

$$\mathcal{S} := \{x \in \mathbb{R}^{n_s} \mid s_i(x) \leq 0, \ i = 1, \ldots, N_s\}.$$

The constraint $s_{i^*}(x) \leq 0$ with $i^* \in \{1, \ldots, N_s\}$ is called *redundant* if its removal does not change $\mathcal{S}$, i.e.

$$\mathcal{S} = \{x \in \mathbb{R}^{n_s} \mid s_i(x) \leq 0, \ i = 1, \ldots, N_s, \ i \neq i^*\}. \qquad \square$$

1.2 Polyhedra

Definition 1.19 (Set collection). S is called a *set collection* (in $\mathbb{R}^{n_S}$) if it is a collection of a finite number of n_s-dimensional sets S_i, i.e.

$$S = \{S_i\}_{i=1}^{N_S},$$

where $\dim(S_i) = n_s$ and $S_i \subseteq \mathbb{R}^{n_s}$ for $i = 1, \ldots, N_S$ with $N_S < \infty$. A set collection is sometimes also referred to as *family of sets*. □

Definition 1.20 (Underlying set). The *underlying set* of a set collection $S = \{S_i\}_{i=1}^{N_S}$ is the point set

$$\underline{S} := \bigcup_{i=1}^{N_S} S_i.$$ □

Usually it is clear from the context if one is talking about the set collection or if it is referred to the underlying set of a set collection, in which case, for simplicity, the same notation is used for both.

Definition 1.21 (Partition). A collection of sets $\{S_i\}_{i=1}^{N_S}$ is a *partition* of a set S if

(i) $S = \cup_{i=1}^{N_S} S_i,$ and

(ii) $S_i \cap S_j = \emptyset$ for all $i \neq j$, where $i, j \in \{1, \ldots, N_S\}$. □

1.2 Polyhedra

The specific class of convex sets called polyhedra play a major role in the framework of this manuscript. As will be shown in the rest of this work, all control and optimization related problems considered here boil down to geometric operations and manipulations on collections of polyhedra.

Most of the following operations and definitions are well known concepts in the computational geometry literature. The interested reader is referred to [Zie95, Grü00] for further detail. Other, more specific definitions (Section 1.2.2), are adopted for the work considered here.

1.2.1 Basic Definitions

The inequality $\cdot \leq \cdot$ in the following is always considered to be component wise.

Definition 1.22 (Hyperplane). A *hyperplane* $\mathcal{P}$ in $\mathbb{R}^{n_x}$ is a set of the form

$$\mathcal{P} = \{x \in \mathbb{R}^{n_x} \mid P^x x = P^0\},$$

where $P^x \in \mathbb{R}^{1 \times n_x}$, $P^x \neq 0'_{n_x}$, $P^0 \in \mathbb{R}$. □

Definition 1.23 (Half-space). A (closed) *half-space* $\mathcal{P}$ in $\mathbb{R}^{n_x}$ is a set of the form

$$\mathcal{P} = \{x \in \mathbb{R}^{n_x} \mid P^x x \leq P^0\},$$

where $P^x \in \mathbb{R}^{1 \times n_x}$, $P^x \neq 0'_{n_x}$, $P^0 \in \mathbb{R}$. □

Definition 1.24 (Polyhedron). A *polyhedron* is the intersection of a finite number of half-spaces. □

Lemma 1.25. A polyhedron $\mathcal{P}$ in $\mathbb{R}^{n_x}$ is a convex set and can always be represented in the form
$$\mathcal{P} = \{x \in \mathbb{R}^{n_x} \mid P^x x \leq P^0\}, \tag{1.1}$$
where $P^x \in \mathbb{R}^{m \times n_x}$, $P^0 \in \mathbb{R}^m$, and $m < \infty$. ∎

Furthermore, if $\|[P^x]_i\|_2 = 1$, where $[P^x]_i$ denotes the i-th row of the matrix P^x in (1.1), the polyhedron $\mathcal{P}$ is called *normalized*.

Definition 1.26 (Face, facet, ridge, edge, vertex). Let $\mathcal{P}$ be a polyhedron in $\mathbb{R}^{n_x}$. A linear inequality $a'x \leq b$ with $a \in \mathbb{R}^{n_x}$ and $b \in \mathbb{R}$ is called *valid* for $\mathcal{P}$ if $a'x \leq b$ holds for all $x \in \mathcal{P}$. A subset $\mathcal{F}$ of $\mathcal{P}$ is called a *face* of $\mathcal{P}$ if it can be represented as

$$\mathcal{F} = \mathcal{P} \cap \{x \in \mathbb{R}^{n_x} \mid a'x = b\},$$

where $a'x \leq b$ is a valid inequality. With $\text{face}^d(\mathcal{P})$ the set of d-dimensional faces of $\mathcal{P}$ is denoted and with $\text{face}_i^d(\mathcal{P})$ its i-th element. (The order is clear from the context.) The faces of the polyhedron $\mathcal{P}$ with dimension 0, 1, $(n_x - 2)$, and $(n_x - 1)$ are called *vertices*, *edges*, *ridges*, and *facets*, respectively. The empty set $\emptyset$ is a face of every polyhedron. □

For the valid inequality $0_{n_x} x \leq 0$, one gets that $\mathcal{P}$ is a face of $\mathcal{P}$. All other faces of $\mathcal{P}$ with $\mathcal{F} \subset \mathcal{P}$ are called *proper faces*.

The dimension of a face $\mathcal{F}$ is the affine dimension of $\mathcal{F}$, i.e. $\dim(\mathcal{F}) = \text{affdim}(\mathcal{F})$.

Definition 1.27 (Minimal representation). Let the polyhedron $\mathcal{P}$ in $\mathbb{R}^{n_x}$ be in the form (1.1). $\mathcal{P}$ is in *minimal representation* if and only if the removal of any row in $P^x x \leq P^0$ would change $\mathcal{P}$, i.e. there are no redundant half-spaces. □

The computational effort required to obtain a minimal representation of a polyhedron, also called *polytope reduction* or *redundancy removal*, is discussed e.g. in [Fuk00b] and generally requires the solution of one linear program for each half-space defining the non-minimal representation of $\mathcal{P}$.

Definition 1.28 (Polytope). A *polytope* is a bounded polyhedron. □

1.2 POLYHEDRA

One of the fundamental theorems for polytopes is the following:

Theorem 1.29 (Polytope representation). $\mathcal{P} \subset \mathbb{R}^{n_x}$ is the convex hull of a finite point set $\mathcal{V} = \{v_1, \ldots, v_{N_v}\}$ for some $v_i \in \mathbb{R}^{n_x}$ with $i = 1, \ldots, N_v$ (V-polytope)

$$\mathcal{P} = \text{conv}\,(\mathcal{V}) = \left\{ x \in \mathbb{R}^{n_x} \;\middle|\; x = \sum_{i=1}^{N_v} \alpha_i v_i,\; 0 \leq \alpha_i \leq 1,\; \sum_{i=1}^{N_v} \alpha_i = 1 \right\}, \quad (1.2)$$

if and only if it is a bounded intersection of half-spaces (H-polytope)

$$\mathcal{P} = \{ x \in \mathbb{R}^{n_x} \mid P^x x \leq P^0 \}, \quad (1.3)$$

for some $P^x \in \mathbb{R}^{m \times n_x}$, $P^0 \in \mathbb{R}^m$, and $m < \infty$. ■

Theorem 1.29 states that a polytope $\mathcal{P}$ can either be expressed in the *vertex representation* (V-representation) as in (1.2) or in its *half-space representation* (H-representation) as in (1.3). The complexity of the translation from one representation to the other is, however, in the worst case exponential depending on the input data and algorithm chosen.

According to Definition 1.8, a polyhedron $\mathcal{P} \subseteq \mathbb{R}^{n_x}$ is *full-dimensional* if it is possible to inscribe an n_x-dimensional ball into $\mathcal{P}$, i.e.

$$\exists x \in \mathbb{R}^{n_x},\; \varepsilon > 0 \quad \text{such that} \quad \mathcal{B}_\varepsilon(x) \subset \mathcal{P}.$$

Otherwise, the polyhedron $\mathcal{P}$ is *lower-dimensional*. A polyhedron is called *empty* if

$$\nexists x \in \mathbb{R}^{n_x} \quad \text{such that} \quad P^x x \leq P^0.$$

Remark 1.30 (Full-dimensional polytope). Throughout this manuscript the focus mostly lies on full-dimensional polytopes. The reason is twofold: (i) numerical difficulties of computation including lower-dimensional polytopes, and, more importantly, (ii) full-dimensional polytopes are sufficient for describing solutions to the problems considered here. For the same reason the MPT toolbox [KGB04] only deals with full-dimensional polytopes. Polyhedra and lower-dimensional polytopes (with the exception of the empty polytope) are not considered. In rare exceptions it is explicitly stated. □

1.2.2 Extensions

As will be seen in the course of this work, one often encounters sets that are disjoint and/or non-convex but that can be represented as the union or collection of a finite number of polyhedra. Therefore, it is useful to define the following (possibly non-standard) mathematical concepts.

Definition 1.31 (Polytope collection). $\mathcal{P}$ is called a *polytope collection* or *P-collection* (in $\mathbb{R}^{n_x}$) if it is a set collection of a finite number of n_x-dimensional polytopes $\mathcal{P}_i$, i.e.

$$\mathcal{P} = \{\mathcal{P}_i\}_{i=1}^{N_P},$$

where $\mathcal{P}_i := \{x \in \mathbb{R}^{n_x} \mid P_i^x x \leq P_i^0\}$, $\dim(\mathcal{P}_i) = n_x$, $i = 1, \ldots, N_P$ with $N_P < \infty$. A P-collection is also referred to as *family of polytopes*. □

Definition 1.32 (Polyhedral partition). A collection of sets $\{\mathcal{P}_i\}_{i=1}^{N_P}$ is a *polyhedral partition* of a set $\mathcal{P}$ if

(i) $\{\mathcal{P}_i\}_{i=1}^{N_P}$ is a partition of $\mathcal{P}$, cf. Definition 1.21, and

(ii) the sets $\bar{\mathcal{P}}_i$ for $i = 1, \ldots, N_P$ are polyhedra, where $\bar{\mathcal{P}}_i$ is the closure of $\mathcal{P}_i$. □

1.2.3 Geometric Operations on Polyhedra

Now some basic operations and functions on polytopes will be defined. Note that although the focus lies on polytopes, most of the operations described here are directly (or with minor modifications) applicable to polyhedra. Additional details on polytope computation can be found in [Zie95, Grü00, Fuk00b]. All operations and functions described in this section are contained in the Multi-Parametric Toolbox (MPT) [KGBC06, KGBM03] for MATLAB®.

Chebyshev Ball

Definition 1.33 (Chebyshev ball). The *Chebyshev ball* of a polytope $\mathcal{P} \subset \mathbb{R}^{n_x}$ is the largest Euclidean ball

$$\mathcal{B}_r(x_c) := \{x \in \mathbb{R}^{n_x} \mid \|x - x_c\|_2 \leq r\},$$

where r is the *Chebyshev radius* and x_c the *Chebyshev center*, such that $\mathcal{B}_r(x_c) \subset \mathcal{P}$. □

For a polytope $\mathcal{P}$ in H-representation (1.3), the center and radius of the Chebyshev ball can be easily found by solving the following linear program [BV04]

$$\max_{x_c, r} \quad r \tag{1.4a}$$

$$\text{subj. to} \quad [P^x]_i x_c + r \|[P^x]_i\|_2 \leq [P^0]_i, \quad i = 1, \ldots, m, \tag{1.4b}$$

where $[P^x]_i$ denotes the i-th row of P^x. If the obtained radius r is zero, then the polytope $\mathcal{P}$ is lower-dimensional; if $r < 0$ then the polytope is empty. Therefore, an answer to the question "is polytope $\mathcal{P}$ full-dimensional or empty?"

is obtained at the expense of only one LP. Furthermore, for full-dimensional polytopes a point x_c that is in the strict interior of $\mathcal{P}$ is obtained. Note that the Chebyshev center x_c in (1.4) is not necessarily unique, e.g. think of $\mathcal{P}$ being a rectangle.

Projection

Definition 1.34 (Projection). Given a set $\mathcal{Q} \subseteq \mathbb{R}^{n_\mathcal{Q}}$ and a set $\mathcal{P} \subseteq \mathbb{R}^{n_\mathcal{P}}$ with $n_\mathcal{Q} \leq n_\mathcal{P} < \infty$, the *(affine) projection* of $\mathcal{P}$ onto $\mathcal{Q}$ is defined as

$$\text{proj}_\mathcal{Q}(\mathcal{P}) := \{q \in \mathcal{Q} \mid \exists p \in \mathcal{P} \text{ with } q = M^p p + M^0\}$$

for some given $M^p \in \mathbb{R}^{n_\mathcal{Q} \times n_\mathcal{P}}$ and $M^0 \in \mathbb{R}^{n_\mathcal{Q}}$. In the literature, projection is also often denoted by $\Pi_\mathcal{Q}(\mathcal{P}) := \text{proj}_\mathcal{Q}(\mathcal{P})$. □

Consider now a special case: given a polytope $\mathcal{P} \subset \mathbb{R}^{n_x + n_y}$ then the *(orthogonal) projection* onto the x-space $\mathbb{R}^{n_x}$ is defined as

$$\text{proj}_x(\mathcal{P}) := \text{proj}_{\mathbb{R}^{n_x}}(\mathcal{P}) = \{x \in \mathbb{R}^{n_x} \mid \exists y \in \mathbb{R}^{n_y} \text{ such that } [x'\ y']' \in \mathcal{P}\}.$$

Current projection methods for polytopes that can operate in general dimensions can be grouped into four classes: Fourier elimination [Cer63, KS90], block elimination [Bal98], vertex based approaches [FLL00], and wrapping-based techniques [JKM04].

Set Difference

The *set difference* of two polytopes $\mathcal{P}$ and $\mathcal{Q}$ is defined by

$$\mathcal{R} = \mathcal{P} \setminus \mathcal{Q} := \{x \mid x \in \mathcal{P}, x \notin \mathcal{Q}\}.$$

One possible way of representing the closure $\bar{\mathcal{R}}$ of $\mathcal{R}$ is by the underlying set $\underline{\mathcal{R}}$ of a P-collection $\{\mathcal{R}_i\}_{i=1}^{N_\mathcal{R}}$, where possibly $\mathcal{R}_i \cap \mathcal{R}_j \neq \emptyset$ for some $i \neq j$. However, whenever it is clear from the context, the notation will be abused and the more compact form will be utilized with the assumption that $\{\mathcal{R}_i\}_{i=1}^{N_\mathcal{R}} = \mathcal{R}$.

The set difference between two P-collections $\mathcal{P}$ and $\mathcal{Q}$ can be computed e.g. via the algorithms described in [BT03, GKBM03, RKM03].

Remark 1.35 (Set difference closure). The set difference of two intersecting polytopes $\mathcal{P}$ and $\mathcal{Q}$ (or any closed sets) is not a closed set. Thus some boundaries (facets) of some $\mathcal{R}_i$ of the P-collection $\mathcal{R} = \mathcal{P} \setminus \mathcal{Q}$ are open, while other borders are closed. Even though it is possible to keep track if faces of $\mathcal{R}_i$ are open or closed, in the algorithms to follow this is not done nor is it performed in the computation in MPT [KGB04, KGBM03, KGBC06], cf. also Remark 1.30.

Henceforth, in the computations and notations only the closure of the sets $\mathcal{R}_i$ are considered. □

Vertex Enumeration

The operation of obtaining the set of vertices vert($\mathcal{P}$) of a polytope $\mathcal{P}$, which is given in H-representation, is called *vertex enumeration*. This operation is the dual to the convex hull computation and the algorithmic implementation is identical to the convex hull computation [Fuk00a, Zie95]. The worst case computational complexity is exponential in the number of input inequalities as there can be exponentially more vertices than half-spaces. An efficient implementation is available in the software package cdd+ by Fukuda [Fuk97].

1.3 Function Terminology

The reader is referred to the excellent textbook by Rockafellar [Roc97] for a general overview on convex analysis.

Definition 1.36 (Affine function). A real-valued function $f : \mathcal{X} \to \mathbb{R}^{n_f}$ with $\mathcal{X} \subseteq \mathbb{R}^{n_x}$ is *affine* if it is of the form

$$f(x) = F^x x + F^0,$$

where $F^x \in \mathbb{R}^{n_f \times n_x}$ and $F^0 \in \mathbb{R}^{n_f}$. □

Definition 1.37 (Piecewise affine function). A real-valued function $f_{\text{PWA}} : \mathcal{X} \to \mathbb{R}^{n_f}$ with $\mathcal{X} \subseteq \mathbb{R}^{n_x}$ is *piecewise affine* (PWA), if $\{\mathcal{X}_i\}_{i=1}^{N_\mathcal{X}}$ is a set partition of $\mathcal{X}$ and

$$f_{\text{PWA}}(x) = F_i^x x + F_i^0 \quad \text{for all} \quad x \in \mathcal{X}_i,$$

where $F_i^x \in \mathbb{R}^{n_f \times n_x}$, $F_i^0 \in \mathbb{R}^{n_f}$, and $i = 1, \ldots, N_\mathcal{X}$. □

Definition 1.38 (Polyhedral PWA function). A real-valued function $f_{\text{PWA}} : \mathcal{X} \to \mathbb{R}^{n_f}$ with $\mathcal{X} \subseteq \mathbb{R}^{n_x}$ is *piecewise affine over polyhedra*, also called *polyhedral piecewise affine* (PPWA), if $\{\mathcal{X}_i\}_{i=1}^{N_\mathcal{X}}$ is a polyhedral partition of $\mathcal{X}$ and

$$f_{\text{PWA}}(x) = F_i^x x + F_i^0 \quad \text{for all} \quad x \in \mathcal{X}_i,$$

where $F_i^x \in \mathbb{R}^{n_f \times n_x}$, $F_i^0 \in \mathbb{R}^{n_f}$, and $i = 1, \ldots, N_\mathcal{X}$. □

Throughout this work, it is usually not distinguished between a piecewise affine function and a polyhedral piecewise affine function due to most partitions here considered being polyhedral partitions. In case of a possible misunderstanding this will be stressed in the context.

1.3 FUNCTION TERMINOLOGY

The definitions for *piecewise quadratic function* (PWQ) and *polyhedral piecewise quadratic function* (PPWQ) follow from Definition 1.37 and 1.38 straightforwardly.

Definition 1.39 (Vector 1-/∞-norm). The *(vector)* 1- and *∞-norm* of $x \in \mathbb{C}^{n_x}$ (or $\mathbb{R}^{n_x}$) is defined as

$$\|x\|_1 := \sum_{i=1}^{n_x} |x_i| \quad \text{and} \quad \|x\|_\infty := \max_{1 \leq i \leq n_x} |x_i|,$$

respectively, where x_i is the *i*-th element of x. □

Definition 1.40 (Matrix 1-/∞-norm). The (induced) matrix 1-norm (or *maximum column-sum norm*) and (induced) matrix ∞-norm (or *maximum row-sum norm*) of $F \in \mathbb{C}^{n_f \times n_x}$ (or $\mathbb{R}^{n_f \times n_x}$) is defined as

$$\|F\|_1 := \sup_{x \neq 0_{n_x}} \frac{\|Fx\|_1}{\|x\|_1} = \max_{1 \leq j \leq n_x} \sum_{i=1}^{n_f} |F_{i,j}| \quad \text{and}$$

$$\|F\|_\infty := \sup_{x \neq 0_{n_x}} \frac{\|Fx\|_\infty}{\|x\|_\infty} = \max_{1 \leq i \leq n_f} \sum_{j=1}^{n_x} |F_{i,j}|,$$

respectively, where $F_{i,j}$ is the (i,j)-th element of F. □

Definition 1.41 (Convex/concave function). A real-valued function $f : \mathcal{X} \rightarrow \mathbb{R}^{n_f}$ is *convex* if its domain $\mathcal{X} \subseteq \mathbb{R}^{n_x}$ is a convex set and

$$\forall x_1, x_2 \in \mathcal{X}, \ 0 \leq \alpha \leq 1 \ \Rightarrow \ f(\alpha x_1 + (1-\alpha)x_2) \leq \alpha f(x) + (1-\alpha)f(x_2),$$

where $\cdot \leq \cdot$ is to be considered component wise. $f(\cdot)$ is *strictly convex* if the last inequality above is replaced by strict inequality. $f(\cdot)$ is *concave* if $-f(\cdot)$ is convex. □

Definition 1.42 (Positive/negative (semi-)definite function). A real-valued function $f : \mathcal{X} \rightarrow \mathbb{R}^{n_f}$ with $\mathcal{X} \subseteq \mathbb{R}^{n_x}$ is *positive definite* if $f(0_{n_x}) = 0_{n_f}$ and $f(x) > 0_{n_f}$ for all $x \neq 0_{n_x}$. $f(\cdot)$ is *positive semi-definite* if $f(0_{n_x}) = 0_{n_f}$ and $f(x) \geq 0_{n_f}$ for all $x \neq 0_{n_x}$. $f(\cdot)$ is called *negative (semi-)definite* if $-f(\cdot)$ is positive (semi-)definite. □

Definition 1.43 (Indefinite function). A real-valued function $f : \mathcal{X} \rightarrow \mathbb{R}^{n_f}$ with $\mathcal{X} \subseteq \mathbb{R}^{n_x}$ is *indefinite* if $f(0_{n_x}) = 0_{n_f}$ and there exists $\bar{x}, \underline{x} \in \mathcal{X}$ such that $f(\bar{x}) > 0_{n_f}$ and $f(\underline{x}) < 0_{n_f}$. □

Stability Theory

The following function classes are commonly used in stability theory [Hah67, GMTT05, Kha96] and are adopted for this work.

Definition 1.44 (K-class function). A real-valued function $\alpha(r)$ with $\alpha : \mathbb{R}_{\geq 0} \to \mathbb{R}_{\geq 0}$ belongs to class K ($\alpha \in K$) if it is continuous and strictly increasing on $0 \leq r \leq \bar{r}$ for some $\bar{r} \in \mathbb{R}_{\geq 0}$, resp. $0 \leq r < \infty$, and if $\alpha(0) = 0$. □

Definition 1.45 ($\tilde{K}$-class function). A real-valued function $\alpha(r)$ with $\alpha : \mathbb{R}_{\geq 0} \to \mathbb{R}_{\geq 0}$ belongs to class $\tilde{K}$ ($\alpha \in \tilde{K}$) if it is continuous, non-decreasing on $0 \leq r \leq \bar{r}$ for some $\bar{r} \in \mathbb{R}_{\geq 0}$, resp. $0 \leq r < \infty$, and if $\alpha(0) = 0$ and $\alpha(r) > 0$ for $r > 0$. □

Definition 1.46 (K_∞-class function). A real-valued function $\alpha(r)$ with $\alpha : \mathbb{R}_{\geq 0} \to \mathbb{R}_{\geq 0}$ belongs to class K_∞ ($\alpha \in K_\infty$) if it is of class K and unbounded, i.e. $\alpha(r) \to +\infty$ as $r \to +\infty$. □

Definition 1.47 (L-class function). A real-valued function $\beta(s)$ with $\beta : \mathbb{R}_{\geq 0} \to \mathbb{R}_{\geq 0}$ belongs to class L ($\beta \in L$) if it is continuous and strictly decreasing on $0 \leq \underline{s} \leq s < \infty$ for some $\underline{s} \in \mathbb{R}_{\geq 0}$ and if $\lim_{s \to \infty} \beta(s) = 0$. □

Definition 1.48 (KL-class function). A real-valued function $\gamma(r,s)$ with $\gamma : \mathbb{R}_{\geq 0} \times \mathbb{R}_{\geq 0} \to \mathbb{R}_{\geq 0}$ belongs to class KL ($\gamma \in KL$), if for each fixed s, $\gamma(\cdot, s)$ is of class K, and for each fixed r, $\gamma(r, \cdot)$ is non-increasing and $\lim_{s \to \infty} \gamma(r,s) = 0$. □

1.4 Constrained Optimization

Mathematical programs have a tremendous impact on various application areas such as engineering, mathematics, physics, and finance. In the context of control or systems theory they play an essential role in almost all advanced techniques, be it in constrained optimal controller synthesis or analysis. Their usage ranges from pure optimal controller computation in order to fulfill particular controller design specifications to system identification, stability and optimality guarantees.

Consider the following generic *mathematical program*, also called a (constrained) *optimization problem*,

$$J^* = \inf_{\theta} \ J(\theta) \tag{1.5a}$$

$$\text{subj. to } \theta \in \mathcal{D}, \tag{1.5b}$$

where $J(\cdot)$ is a real-valued *cost function* (also called *objective function*) defined over the domain $\mathcal{D} \subseteq \mathbb{R}^{n_\theta}$, J^* is the *optimal value*, $\theta \in \mathbb{R}^{n_\theta}$ denotes the *optimization variable*, and a vector θ^* of (1.5) such that $J(\theta^*) = J^*$ is called solution or *optimizer* of the program (1.5).

1.4 Constrained Optimization

Mathematical programming has been well studied and a vast amount of literature exists [BS79, BTN01, Ber95, BV04, BEFB94, Sch86, Flo95]. In their full generality, however, optimization problems are very hard to solve. Fortunately, for the important and broad subclass of convex optimization problems, efficient solvers exist.

Definition 1.49 (Convex optimization problem). The mathematical program (1.5) is called a *convex optimization problem* if both the cost function $J(\cdot)$ and the set $\mathcal{D}$ are convex. □

For even more structured optimization problems, as detailed below, many dedicated solvers with polynomial complexity exist and scale well for large problems. *Polynomial complexity* means that the effort needed to find a solution to the optimization problem or to find a certificate for the non-existence of a solution grows polynomially with the data size[1] of the optimization problem and the required accuracy of the solution, cf. [BTN01].

Here only a rough overview of the most important standard programs used in the following chapters are given. Note that many other equivalent 'standard' formulations are described in the literature. The different optimization problems are ordered by the complexity involved in solving them.

Remark 1.50 (Infimum/minimum problem). For the optimization problems considered in this work the prerequisites of the Weierstrass' theorem [Ber95] are always fulfilled and thus the existence of an optimizer is guaranteed (if the optimization problems are solvable). As a consequence, the infimum in (1.5a) is replaced with minimum. □

Linear Program

A *linear program* (LP) can be given in the form

$$J^* = \min_{\theta} \quad j'\theta \tag{1.6a}$$

$$\text{subj. to} \quad D^\theta \theta \leq D^0, \tag{1.6b}$$

where the optimization variable $\theta \in \mathbb{R}^{n_\theta}$, $j \in \mathbb{R}^{n_\theta}$, $D^\theta \in \mathbb{R}^{n_D \times n_\theta}$, and $D^0 \in \mathbb{R}^{n_D}$.

Theoretically every LP is solvable in polynomial time by both the ellipsoid method of Khachiyan [Kha79, Sch86] and various interior point methods [Kar84, RTV97]. The two fundamentally different algorithms of practical relevance are the simplex and the interior point method. A practical algorithm to solve an LP with n_θ variables and n_D constraints requires roughly $O(n_\theta^3 n_D^{1/2} + n_\theta^2 n_D^{3/2})$ operations [dH94].

---
[1] The 'size' of an optimization problem is largely influenced by the dimension of the optimization variable and the number of constraints.

Due the existence of very efficient solvers for LPs and compared to the more involved operations considered in this manuscript, we can view an LP as a very simple operation. For comparison, in some instances throughout the following chapters the complexity of an algorithm will be expressed in terms of the number of LPs one needs to solve.

Quadratic Program

Similarly, a convex *quadratic program* (QP) can be given in the commonly used form

$$J^* = \min_{\theta} \quad \tfrac{1}{2}\theta'(J^{\theta\theta})\theta + j'\theta \tag{1.7a}$$

$$\text{subj. to} \quad D^\theta \theta \leq D^0, \tag{1.7b}$$

where the optimization variable $\theta \in \mathbb{R}^{n_\theta}$, $J^{\theta\theta} \in \mathbb{R}^{n_\theta \times n_\theta}$ with $J^{\theta\theta} = (J^{\theta\theta})' \succeq 0$, $j \in \mathbb{R}^{n_\theta}$, $D^\theta \in \mathbb{R}^{n_D \times n_\theta}$, and $D^0 \in \mathbb{R}^{n_D}$.

A quadratic program can be solved with roughly the same efficiency as a linear program, but on average the solvers are approximately five times slower than LP solvers [Neu04].

Semidefinite Program and Linear Matrix Inequality

A *semidefinite program* (SDP) is a convex optimization problem that can be expressed in the standard (dual) form

$$J^* = \min_{\theta} \quad j'\theta \tag{1.8a}$$

$$\text{subj. to} \quad D^\theta \theta = D^0, \tag{1.8b}$$

$$M_0 + \sum_{i=1}^{n_\theta} \theta_i M_i \preceq 0, \tag{1.8c}$$

where the optimization variable $\theta \in \mathbb{R}^{n_\theta}$, $j \in \mathbb{R}^{n_\theta}$, $D^\theta \in \mathbb{R}^{n_D \times n_\theta}$, $D^0 \in \mathbb{R}^{n_D}$, and $M_i = M_i' \in \mathbb{R}^{n_M \times n_M}$ for $i = 0, \ldots, n_\theta$.

The inequality (1.8c) is usually referred to as *linear matrix inequality* (LMI) [BEFB94]. It is apparent from (1.8c) that LMIs imply a feasible set of parameters and are not, as it is often mistakenly perceived, an optimization problem.

It is illustrated in [BEFB94] that very many convex problems can be posed as an SDP, such as $\mathcal{H}_\infty$-control or the search for a Lyapunov function of a closed-loop system.

Although SDPs can be solved in polynomial time, using interior point methods, current solvers are considerably slower (roughly one order of magnitude) than current LP or QP solvers. Note, that LPs and QPs are a subclass of SDPs.

Mixed-Integer Linear Program

A different type of mathematical program is the following *mixed-integer linear program* (MILP) [Sch86, Flo95]

$$J^* = \min_{\theta} \quad j'\theta \tag{1.9a}$$

$$\text{subj. to} \quad D^\theta \theta \leq D^0, \tag{1.9b}$$

where the optimization variable is partitioned as

$$\theta = \begin{bmatrix} \theta_\mathbb{R} \\ \theta_\mathbb{B} \end{bmatrix} \quad \text{with} \quad \theta_\mathbb{R} \in \mathbb{R}^{n_{\theta,R}}, \quad \theta_\mathbb{B} \in \{0,1\}^{n_{\theta,B}},$$

and $n_\theta = n_{\theta,R} + n_{\theta,B}$. $j \in \mathbb{R}^{n_\theta}$, $D^\theta \in \mathbb{R}^{n_D \times n_\theta}$, and $D^0 \in \mathbb{R}^{n_D}$.

Even though the problem looks structurally simple due to the linear cost function and linear set of constraints and is thus similar to an LP, it is non-convex and belongs to the class of NP-hard problems, i.e. the problem solving time grows in the worst case exponentially with the number $n_{\theta,B}$ of binary variables. However, a number of successful algorithms and solvers exist.

1.5 Multi-Parametric Programming

Most constrained control and analysis algorithms described throughout this manuscript use the underlying core concept of multi-parametric programming, which is shortly explained in the following. The interested reader is referred to [BGK+82] for the general mathematical concept and properties of multi-parametric programs and to [Bor03, Tøn00, Bao05, Jon05] and the references therein for algorithms to compute their solutions and its usage in the framework of control.

Consider the nonlinear optimization problem

$$J^*(x) = \min_{\theta} \quad J(\theta, x) \tag{1.10a}$$

$$\text{subj. to} \quad \begin{bmatrix} \theta \\ x \end{bmatrix} \in \mathcal{D}, \tag{1.10b}$$

where $J(\cdot, \cdot)$ is a real-valued *cost function* defined over the domain $\mathcal{D} \subseteq \mathbb{R}^{n_\theta + n_x}$, $J^*(\cdot)$ is the *value function*, $x \in \mathbb{R}^{n_x}$ denotes a *parameter vector* (in the following chapters mostly the system state), and $\theta \in \mathbb{R}^{n_\theta}$ denotes the *optimization variable* (in the following chapters it usually includes the control input variable u).

The optimization problem (1.10) is commonly referred to as *parametric* or *multi[2]-parametric (nonlinear) program* (mp-NLP). The aim is to obtain the value $J^*(x)$ and optimizer $\theta^*(x)$ or set of optimizers

---

[2] The word 'multi' denotes the fact, that the parameter x is not a scalar but a vector.

$$\Theta^*(x) = \{\theta \in \mathbb{R}^{n_\theta} \mid J^*(x) = J(\theta, x),\ [\theta'\ x']' \in \mathcal{D}\}$$

for a whole range of feasible parameters $x \in \mathcal{X}$ and not only for a fixed parameter as presented in Section 1.4. In its generality $\theta^*(x) \in \Theta^*(x)$, where $\Theta^* : \mathcal{X} \to 2^{\mathbb{R}^{n_\theta}}$ denotes the point-to-set mapping from x to the set of optimizers. (The powerset $2^{\mathbb{R}^{n_\theta}}$ is the set of all subsets of $\mathbb{R}^{n_\theta}$.)

Definition 1.51 (Feasibility). The optimization problem (1.10) is called *feasible* if $\mathcal{D} \neq \emptyset$. The point $x \in \mathbb{R}^{n_x}$ (resp. $\theta \in \mathbb{R}^{n_\theta}$) is called *feasible* if and only if there exists a $\theta \in \mathbb{R}^{n_\theta}$ (resp. $\exists x \in \mathbb{R}^{n_x}$) such that $[\theta'\ x']' \in \mathcal{D}$. □

It is intuitively clear, that depending on the form and properties of the cost function $J(\cdot, \cdot)$ and the structure of the constraint (1.10b), the optimizer $\theta^*(\cdot)$ and the value function $J^*(\cdot)$ may vary continuously, smoothly, or even discontinuously with x. One can distinguish the two most important special classes of multi-parametric programs for control purposes, i.e. if the constraints are linear and the cost function is either quadratic or linear in the optimizer, then the optimization problem (1.10) is called a *multi-parametric quadratic program* (mp-QP) or a *multi-parametric linear program* (mp-LP), respectively.

For the work at hand the main interest lies in the special case of multi-parametric linear programs, which will be shortly described in the following. The reader is referred to [Bor03, Tøn00] for a detailed description of mp-QPs.

Multi-Parametric Linear Programming

Consider the *multi-parametric linear program* (mp-LP)

$$J^*(x) = \min_\theta\quad j'\theta \tag{1.11a}$$
$$\text{subj. to}\quad D^\theta \theta \leq D^0 + D^x x, \tag{1.11b}$$

where $x \in \mathbb{R}^{n_x}$ denotes the parameter vector, $\theta \in \mathbb{R}^{n_\theta}$ denotes the optimization variable, $J^*(\cdot)$ is the value function, $j \in \mathbb{R}^{n_\theta}$, $D^\theta \in \mathbb{R}^{n_D \times n_\theta}$, $D^0 \in \mathbb{R}^{n_D}$, and $D^x \in \mathbb{R}^{n_D \times n_x}$.

The case when the parameter enters linearly the cost function and is absent in the linear constraints, i.e.

$$J(\theta, x) = (J^{\theta x} x + j)' \theta \quad \text{with}\quad J^{\theta x} \in \mathbb{R}^{n_\theta \times n_x}\ \text{and}\ D^\theta \theta \leq D^0,$$

can be reformulated into the above mentioned form (1.11) via the dual form of the LP. The solution of a mixed setup, where a set of elements of the parameter vector x enters linearly the cost function and a different set of elements of the parameter vector enters linearly the constraint set, was solved in [BBM05].

1.5 Multi-Parametric Programming

The following theorem, proved in [Gal95], describes the properties of the mp-LP solution.

Theorem 1.52 (mp-LP solution properties). Consider the multi-parametric linear program (1.11). The set of feasible parameters $\mathcal{X}$ is a closed polyhedron in $\mathbb{R}^{n_x}$. The function $J^*(\cdot)$ is convex and piecewise affine over $\mathcal{X}$.
If the optimizer $\Theta^* : \mathcal{X} \to 2^{\mathbb{R}^{n_\theta}}$ is a unique single-valued function $\theta^* : \mathcal{X} \to \mathbb{R}^{n_\theta}$ for all $x \in \mathcal{X}$ then $\theta^*(\cdot)$ is continuous and polyhedral piecewise affine. Otherwise, it is always possible to define a continuous and polyhedral piecewise affine optimizer function $\theta^*(x)$ for all $\mathcal{X}$. ∎

The original work to solve parametric linear programming problems goes back to the Master's thesis of Orchard-Hays of 1952, published in [OH55] and Saaty and Gass [SG54]. A survey of the early work in the field of parametric linear programming is given in [Gal80]. Gal and Nedoma [Gal95] propose to solve (1.11) by a simplex method based approach where all optimal bases of the problem are enumerated. Geometric algorithms based on parameter space exploration ideas are detailed in [Bor03, Tøn00, Bao05].

The main problem for all above mentioned mp-LP algorithms is the possibility of dual degeneracy of the underlying LPs, i.e. when the optimizer is not unique. A non-unique optimizer can result in an overlapping partitioning of the parameter space. This artifact together with the possibility of a discontinuous optimizer might lead to severe effects such as chattering of the control action when using the mp-LP solution (or directly the underlying LP solution) as a control map from the system state to the control input on the system to be controlled. These issues can for example be remedied by using an mp-LP algorithm based on lexicographic perturbation as proposed in [Jon05, JKM05]. An alternative way to handle dual degeneracies in an mp-LP is described in [STJ05], where in each dual degenerate situation an additional mp-QP is solved in order to obtain a unique solution for the overall mp-LP problem. Thus a unique, continuous, non-overlapping, and polyhedral piecewise affine optimizer $\theta^*(x)$ for all feasible $x \in \mathcal{X}$ can be obtained.

2
SYSTEMS AND CONTROL THEORY

In this chapter some basic systems theoretical definitions and theorems are provided, which are used throughout the text.

2.1 Systems Formulation

We consider a discrete-time nonlinear system described by the following *state-update equation*

$$x(t+1) = f(x(t), u(t), t), \qquad (2.1a)$$

where the time $t \in \mathbb{T} \subseteq \mathbb{Z}$ and the system is subject to the following state and input *constraints*

$$x(t) \in \mathbb{X} \subseteq \mathbb{R}^{n_x} \quad \text{and} \quad u(t) \in \mathbb{U} \subseteq \mathbb{R}^{n_u}, \qquad (2.1b)$$

respectively. Note that the constraints (2.1b) can be generalized to a mixed state-input constraint of the form

$$[x(t)' \ u(t)']' \in \mathbb{D} \subseteq \mathbb{R}^{n_x + n_u} \qquad (2.1b')$$

and that no (continuity) assumption is made for the function $f(\cdot, \cdot, \cdot)$.

2.2 Invariance and Stability

Pre-specified (or system imposed) constraints on the system state, input, state-input, and time-axis are denoted with $\mathbb{X}$, $\mathbb{U}$, $\mathbb{D}$, and $\mathbb{T}$, respectively. Most often, however, control problems are not solvable for the whole domain $\mathbb{X} \times \mathbb{U}$ or $\mathbb{D}$. For this purpose the calligraphic letters

$$\mathcal{X} \subseteq \mathbb{X}, \quad \mathcal{U} \subseteq \mathbb{U}, \quad \text{or} \quad \mathcal{D} \subseteq \mathbb{D}$$

will be used to specify the generic sets of states, inputs, and mixed state-inputs, respectively, for which the control problem is solvable or of particular interest.

A *shorthand notation* for system (2.1) is given by

$$x^+ = f(x, u, t), \qquad (2.1a'')$$

where $x^+ := x(t+1)$ denotes the successor state of $x := x(t)$.

Definition 2.1 (Autonomous system). The system (2.1) is called *autonomous* if it is of the special form

$$x(t+1) = f(x(t), t), \qquad (2.2)$$

where $x(t) \in \mathbb{X} \subseteq \mathbb{R}^{n_x}$ for all $t \in \mathbb{T} \subseteq \mathbb{Z}_{\geq 0}$. $\square$

Clearly, a closed-loop system with state-update function

$$f^{\text{CL}}(x(t), t) := f(x(t), \mu(x(t), t), t) \quad \text{under control} \quad u(t) = \mu(x(t), t)$$

is autonomous.

2.2 Invariance and Stability

Most of the following concepts from systems theory are well known and can be found in the literature, cf. e.g. [Fre65, GSD05, Hah67, Lya92, Kha96, Vid93, GT91, Ker00]. Others are slightly modified for the framework considered here.

Definition 2.2 ((Positively) Invariant set $\mathcal{O}$). A set $\mathcal{O} \subseteq \mathbb{X}$ is said to be *invariant* with respect to the autonomous system (2.2) if and only if

$$x(t_0) \in \mathcal{O}, \ (t_0 \in \mathbb{T}) \quad \Rightarrow \quad x(t) \in \mathcal{O}, \quad \forall t \in \mathbb{T}.$$

Moreover, a set $\mathcal{O} \subseteq \mathbb{X}$ is said to be *positively invariant* if and only if

$$x(t_0) \in \mathcal{O}, \ (t_0 \in \mathbb{T}) \quad \Rightarrow \quad x(t) \in \mathcal{O}, \quad \forall t \geq t_0, \ t \in \mathbb{T}. \quad \square$$

Note that the union of two (positively) invariant sets is (positively) invariant. However, in general, the intersection is not invariant [Ker00].

Definition 2.3 (Minimal/maximal positively invariant set). A set $\mathcal{O}_{\min} \subseteq \mathbb{X}$ is said to be a *minimal positively invariant set* with respect to the autonomous system (2.2) if and only if $\mathcal{O}_{\min}$ is a positively invariant set and $\mathcal{O}_{\min}$ is a subset of all positively invariant sets.

A set $\mathcal{O}_{\max} \subseteq \mathbb{X}$ is said to be a *maximal positively invariant set* with respect to the autonomous system (2.2) if and only if $\mathcal{O}_{\max}$ is a positively invariant set and $\mathcal{O}_{\max}$ contains all positively invariant sets. □

If the autonomous system (2.2) has disconnected positively invariant sets, e.g. in the case of multiple equilibria, then $\mathcal{O}_{\min} = \emptyset$. In the 'desired' case in control where the closed-loop autonomous system only has one equilibrium point, i.e. for example the origin, it follows $\mathcal{O}_{\min} = \{0_{n_x}\}$.

Definition 2.4 (Input admissible set $\mathcal{X}^\mu$). Consider system (2.1). For a given state feedback control law $u(t) = \mu(x(t), t)$ with $t \in \mathbb{T}$ and a given subset of states $\mathcal{X} \subseteq \mathbb{X}$ the *input admissible set* is defined as

$$\mathcal{X}^\mu := \{x(t) \in \mathcal{X} \mid \mu(x(t), t) \in \mathbb{U}\}.$$

□

In the case of a specified control law $\mu(\cdot, \cdot)$, the state constraint $x \in \mathbb{X}$ in (2.1b) can be replaced by $x \in \mathbb{X}^\mu$ without changing or truncating the feasible state-space.

Definition 2.5 ((Maximal) control invariant set $\mathcal{C}_{\max}$). The set $\mathcal{C} \subseteq \mathbb{X}$ is *control invariant* for system (2.1) if and only if

$$x(t_0) \in \mathcal{C}, \ (t_0 \in \mathbb{T}) \ \Rightarrow \ \exists u(t) \in \mathbb{U} : f(x(t), u(t), t) \in \mathcal{C}, \ \forall t \geq t_0, \ t \in \mathbb{T}.$$

The set $\mathcal{C}_{\max} \subseteq \mathbb{X}$ is the *maximal control invariant set* for system (2.1) if it is control invariant and contains all control invariant sets in $\mathbb{X}$. □

It follows immediately that the union of two control invariant sets is control invariant. Whereas, in general, the intersection of two control invariant sets is not necessarily control invariant. Moreover, even though it follows that for any $x(t) \in \mathcal{C}$ there exists a control strategy to keep the trajectory in $\mathcal{C}$, this does not imply asymptotic stability of the closed-loop system nor attractivity of some subset $\mathcal{X} \subset \mathcal{C}$ to the equilibrium point. These concepts are introduced in the following.

Definition 2.6 (Attractivity). The set $\mathcal{S} \subseteq \mathbb{X}$ is *attractive* with respect to the autonomous system (2.2) if there exists a set $\mathcal{A} \subseteq \mathbb{X}$ with $\mathcal{A} \neq \emptyset$ of initial states such that

$$x(t_0) \in \mathcal{A}, \ (t_0 \in \mathbb{Z}_{\geq 0}) \ \Rightarrow \ \lim_{t \to \infty} \|x(t) - \mathcal{S}\| = 0.$$

The set $\mathcal{A}$ is called a *region of attraction* to $\mathcal{S}$. $\mathcal{A}_{\max} \subseteq \mathbb{X}$ is the *maximal region of attraction* to $\mathcal{S}$ if it is a region of attraction and contains all regions of attractions in $\mathbb{X}$ to $\mathcal{S}$. □

2.2 Invariance and Stability

The *point-to-set distance* is defined by

$$\|x - \mathcal{S}\| := \inf_{s \in \mathcal{S}} \|x - s\|,$$

where $\|\cdot\|$ is any vector norm; and usually $\mathcal{S}$ is the equilibrium point x_{ep} of the autonomous system (2.2), i.e. $\mathcal{S} = \{x_{\text{ep}}\}$.

Definition 2.7 ((Lyapunov) stability). The equilibrium point $x_{\text{ep}} \in \mathcal{X} \subseteq \mathbb{X}$ of the autonomous system (2.2) is *(Lyapunov) stable* in $\mathcal{X}$ if for every $\varepsilon > 0$, there exists a $\delta(\varepsilon, t_0) > 0$ such that, for every $x(t_0) \in \mathcal{X}$ and $t_0 \in \mathbb{Z}_{\geq 0}$

$$\|x(t_0) - x_{\text{ep}}\| < \delta(\varepsilon, t_0) \quad \Rightarrow \quad \|x(t) - x_{\text{ep}}\| < \varepsilon \quad \forall t \geq t_0. \qquad \Box$$

Definition 2.8 (Asymptotic stability). The equilibrium point $x_{\text{eq}} \in \mathcal{X} \subseteq \mathbb{X}$ of the autonomous system (2.2) is *asymptotically stable* in $\mathcal{X}$ if it is stable and attractive. $\Box$

Definition 2.9 (Exponential stability). The equilibrium point $x_{\text{ep}} \in \mathcal{X} \subseteq \mathbb{X}$ of the autonomous system (2.2) is *exponentially stable* in $\mathcal{X}$ if there exist constants $\alpha > 0$ and $\gamma \in (0,1)$ such that, for every $x(t_0) \in \mathcal{X}$ and $t_0 \in \mathbb{Z}_{\geq 0}$

$$x(t_0) \in \mathcal{X} \quad \Rightarrow \quad \|x(t) - x_{\text{ep}}\| \leq \alpha \|x(t_0) - x_{\text{ep}}\| \gamma^t \quad \forall t \geq t_0. \qquad \Box$$

Definition 2.10 (Lyapunov function). Consider the autonomous system (2.2) with equilibrium point $x_{\text{ep}} = \mathbb{0}_{n_x}$. A positive definite function $V : \mathcal{X} \to \mathbb{R}_{\geq 0}$ with $\mathcal{X} \subseteq \mathbb{X}$, $\mathcal{X}$ positively invariant, $V(\mathbb{0}_{n_x}) = 0$, and $V(f(x(t), t)) - V(x(t)) \leq 0$ for all $x(t) \in \mathcal{X}$ and $t \in \mathbb{T}$ is called *Lyapunov function* for the system (2.2) on $\mathcal{X}$. $\Box$

It seems that the issue of stability for *discontinuous* discrete-time nonlinear systems in control, which are considered in this work, only recently found proper attention in the literature. Earlier results on stability for discrete-time systems, such as e.g. [Fre65], usually assume (Lipschitz) continuity of the Lyapunov function $V(\cdot)$ and/or the state-update function $f(\cdot)$. Even though [Vid93] stated a theorem for asymptotic and exponential stability for discrete-time systems, they are given without any proof. Moreover, the result for exponential stability in [Vid93] is unnecessarily conservative. A good summary on stability for discontinuous discrete-time systems is presented in [GSD05]. Here, the reader is referred to the excellent overview on stability for *non-smooth systems* by Lazar [Laz06]. The comparatively general result on asymptotic and exponential stability are repeated for completeness in the following theorem.

Theorem 2.11 (Asymptotic/exponential stability). Let $\mathcal{X} \subseteq \mathbb{X}$ be a bounded positively invariant set for the autonomous system (2.2) that contains a neighborhood $\mathcal{N}(x_{\text{ep}})$ of the equilibrium $x_{\text{ep}} = \mathbb{0}_{n_x}$, and let $\underline{\alpha}(\cdot), \overline{\alpha}(\cdot)$, and $\beta(\cdot)$ be K-class functions, cf. Definition 1.44.

If there exists a non-negative function $V : \mathcal{X} \to \mathbb{R}_{\geq 0}$ with $V(\mathbb{0}_{n_x}) = 0$ such that

$$V(x) \geq \underline{\alpha}(\|x\|), \quad x \in \mathcal{X}, \tag{2.3a}$$
$$V(x) \leq \overline{\alpha}(\|x\|), \quad x \in \mathcal{N}(x_{\text{ep}}), \tag{2.3b}$$
$$\Delta V(x) := V(f(x)) - V(x) \leq -\beta(\|x\|), \quad x \in \mathcal{X}, \tag{2.3c}$$

then the following results holds:

(a) The equilibrium point $\mathbb{0}_{n_x}$ is *asymptotically stable* in the Lyapunov sense in $\mathcal{X}$.

(b) If $\underline{\alpha}(\|x\|) := \underline{a}\|x\|^\gamma$, $\overline{\alpha}(\|x\|) := \overline{a}\|x\|^\gamma$, and $\beta(\|x\|) := b\|x\|^\gamma$ for some positive constants $\underline{a}, \overline{a}, b, \gamma > 0$ then the equilibrium point $\mathbb{0}_{n_x}$ is *locally exponentially stable*. Moreover, if the inequality (2.3b) holds for $\mathcal{N}(x_{\text{ep}}) = \mathcal{X}$, then the equilibrium point $\mathbb{0}_{n_x}$ is *exponentially stable* in the Lyapunov sense in $\mathcal{X}$. ∎

It is crucial for later results to point out that Theorem 2.11 permits both $V(\cdot)$ and $f(\cdot)$ to be discontinuous for any $x \neq \mathbb{0}_{n_x}$. Therefore, continuity of $V(\cdot)$ and $f(\cdot)$ are only required *at* the equilibrium point $\mathbb{0}_{n_x}$ and not necessarily for a neighborhood around $\mathbb{0}_{n_x}$.

Artstein [Art83] showed for continuous-time systems that the existence of a global asymptotic stabilizing control law is equivalent to the existence of a control Lyapunov function. Many of the ideas were transferred to discrete-time systems in [KT03]. Thus control Lyapunov functions [KT03, GMTT05, Son83, Art83, KKK95] form an interesting and important means of designing stabilizing control laws.

Definition 2.12 (Control Lyapunov function). Consider system (2.1), a function $V_{\text{CLF}} : \mathbb{R}^{n_x} \to \mathbb{R}_{\geq 0}$, and a set $\mathcal{X} \subseteq \mathbb{X}$ with $\mathbb{0}_{n_x} \in \text{int}(\mathcal{X})$. $V_{\text{CLF}}(\cdot)$ is a *control Lyapunov function* (CLF) on $\mathcal{X}$ if the following conditions hold:

(a) There exists class K_∞ functions $\underline{\alpha}(\cdot)$ and $\overline{\alpha}(\cdot)$ such that $\underline{\alpha}(\|x\|) \leq V_{\text{CLF}}(x) \leq \overline{\alpha}(\|x\|)$ for all $x \in \mathcal{X}$, and

(b) There exists a continuous positive definite function $\beta : \mathbb{R}^{n_x} \to \mathbb{R}_{\geq 0}$ such that for each $x \in \mathcal{X}$ there exists a $u \in \mathbb{U}$ with $V_{\text{CLF}}(f(x,u)) - V_{\text{CLF}}(x) \leq -\beta(x)$. □

2.2 Invariance and Stability

Note that a (possibly discontinuous) asymptotically stable closed-loop system has some robustness with respect to measurement noise d_x and additive disturbance w, i.e. $x^+ = f(x + d_x, u) + w$, if and only if there exists a continuous Lyapunov function for the system [Kel02, KT02].

3
RECEDING HORIZON CONTROL

The *Receding Horizon Control* (RHC) paradigm, often also interchangeably called *Model Predictive Control* (MPC) or *Model Based Predictive Control* (MBPC), is the most successful modern optimal control strategy for constrained systems in practice and numerous theoretical as well as application oriented publications are available. To name a few: in the area of textbooks there are [Mac02, GSD05, Ros03, KC01, CB04] as well as the survey papers [MRRS00, May01b, BM99b, GPM89, QB97, Raw00, ML99]. The reader is referred to [Mac02, GSD05] for a good introduction to most of the ideas.

Although the basic idea of receding horizon control was already indicated by the theoretical work of Propoi [Pro63] in 1963 it did not gain much attention until the mid-1970s, when Richalet *et al.* [RRTP76] proposed a technique called Model Predictive Heuristic Control (MPHC). Shortly after, Cutler and Ramaker [CR80] presented the predictive control algorithm called Dynamic Matrix Control (DMC) which has lead to a huge success in the petro-chemical industry. A vast variety of different names and methodologies followed, such as Quadratic Dynamic Matrix Control (QDMC), Adaptive Predictive Control (APC), Generalized Predictive Control (GPC), Sequential Open Loop Optimization (SOLO), and others.

3.1 BASIC IDEA

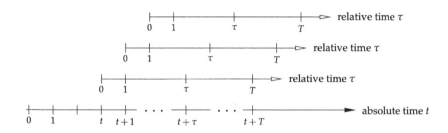

Fig. 3.1: Relation of the absolute time axis and the shifting relative prediction time axis.

While the proposed algorithms are seemingly different, they all share the same structural features: a model of the plant, the receding horizon idea, and an optimization procedure to obtain the control action by predicting the system's evolution.

3.1 Basic Idea

The basic idea of receding horizon control is well established for general nonlinear systems and the only extensions performed are in order to guarantee closed-loop stability, feasibility for all time, optimal performance, and robustness.

The idea is the following: on the basis of a *model*[1] of the system

$$x(\tau + 1) = f(x(\tau), u(\tau)), \quad \text{where} \quad \begin{bmatrix} x(\tau) \\ u(\tau) \end{bmatrix} \in \mathbb{X} \times \mathbb{U} \text{ and } \tau \in [0, T], \quad (3.1)$$

and a measurement $x_t =: x(0)$ of the system's state at sampling time t, i.e. $\tau = 0$ (cf. Figure 3.1), an optimization problem is formulated that *predicts* over a finite horizon T the optimal evolution of the state according to some optimality criterion. The result of the optimization procedure is an optimal control input sequence

$$U_T^*(x_t) := \{u(0), u(1), \ldots, u(T-1)\}.$$

Even though a whole sequence of inputs is at hand, only the first element $u(0)$ is applied to the system. In order to compensate for possible modeling errors or disturbances acting on the system, a new state measurement x_{t+1} is taken at the next sampling instance $t + 1$ and the whole procedure is repeated. This introduces feedback to the system.

---

[1] The model of the system does not necessarily need to be a state space difference equation as considered here but any means to predict the evolution of the system over a finite horizon. This could, for example, be a finite impulse response model, a neural network, or any oracle.

Remark 3.1 (Absolute and relative time axis). Note that τ has the meaning of a time instance on the relative prediction time axis that shifts with every new sampling time, cf. Figure 3.1. A commonly used, more precise, notation would be $x(t + \tau|t)$, i.e. $x(t + \tau|t)$ is the predicted state vector at the absolute time instance $t + \tau$ based on the measured state vector x_t taken at the absolute time instance t and a given system prediction model. However, for brevity reasons this will be omitted as the simpler notation is clear in its context. □

From the above explanations it is clear that a fixed prediction horizon is shifted or *receded* over time, hence its name, receding horizon control. The procedure of this *on-line* optimal control technique is summarized in the following algorithm.

Algorithm 3.2 (On-line receding horizon control)
1. measure the state x_t at time instance t
2. obtain $U_T^*(x_t)$ by solving an optimization problem over the horizon T
   IF $U_T^*(x_t) = \emptyset$ **THEN** problem infeasible **STOP**
3. apply the first element $u(0)$ of $U_T^*(x_t)$ to the system
4. wait for the new sampling time $t + 1$, goto (1.)

3.2 Closed-form Solution to Receding Horizon Control

As mentioned above, the on-line receding horizon control strategy is and has been very successful in the application area of chemical processes, where the usual time constants of the systems range from hours to days, or even weeks. Therefore, the time-span required, within which the overall optimization problem needs to be solved, is practically 'unlimited'.

In contrast, fast systems that need to be sampled and controlled in the range of milli- or microseconds limit the complexity of the overall optimization problem in Step 2 of Algorithm 3.2. This usually leads either to a setup of the optimization problem where only a very short prediction horizon can be considered, a simpler prediction model is taken into account, or, in the worst case, on-line receding horizon control cannot be applied to the particular system at all.

Only recently, a possible remedy for this 'deficiency' or 'lack of time' was proposed [BMDP02, Bor03]: unlike before, the state measurement $x_t = x(0)$ in the formulation of the overall optimization problem is not considered as one fixed vector but as a free parameter and thus the on-line optimization procedure can be performed at once *off-line* for all $x_t \in \mathbb{X}$. The solution to the resulting *multi-parametric optimization program*, cf. Section 1.5, is a closed-form (possibly

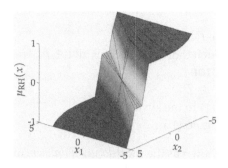 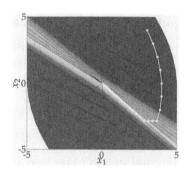

Fig. 3.2: Receding horizon control closed-form solution $\mu_{RH}(\cdot)$ for a simple constrained double integrator system.

Fig. 3.3: State-space partition $\{\mathcal{P}_i\}_{i=1}^{N_\mathcal{P}}$ of the closed-form solution $\mu_{RH}(\cdot)$ including a state trajectory starting at $x(0) = [3, 4]'$. Coloring corresponds to Figure 3.2.

nonlinear) time-varying piecewise *state feedback control law* μ, which is a function that returns the optimal control input, i.e.

$$u(\tau) = \mu(x_t, \tau), \quad (3.2)$$

where $\tau \in \{0, \ldots, T-1\}$ and is equivalent to solving for $U_T^*(x_t)$ for all $x_t \in \mathbb{X}$.

As mentioned above, according to the receding horizon control policy, only the first element $u(0)$ is taken into account and thus (3.2) simplifies to the closed-form time-invariant piecewise state feedback control law

$$\mu_{RH}(x_t) := \mu(x_t, 0) = \mu_i(x_t), \quad \text{if} \quad x_t \in \mathcal{P}_i, \quad (3.2')$$

where $i = 1, \ldots, N_\mathcal{P}$ and $x_t \in \mathbb{X} \subseteq \mathbb{R}^{n_x}$ denotes the measured state of the controlled system at time $t \geq 0$. $\mu_i(\cdot) \in \mathbb{U} \subseteq \mathbb{R}^{n_u}$ are in general different nonlinear control functions (or oracles) over the sets $\mathcal{P}_i$ and are possibly overlapping, i.e. there exists $\mathcal{P}_i$ and $\mathcal{P}_j$ with $i \neq j$ such that $\mathcal{P}_i \cap \mathcal{P}_j$ is non-empty.

The state feedback control law (3.2') is also referred to as the *optimal lookup table* or the *explicit solution* of the receding horizon controller. See Figure 3.2 and 3.3 for a simple example.

In an on-line application the control action is given by

$$u(t) = \mu_{RH}(x(t)),$$

where now and in the following chapters $x(t)$ simply denotes the system's state at time t. In order to evaluate the control input one needs to identify the state space region $\mathcal{P}_i$ in which the measured state $x(t)$ lies at the sampling instance t, i.e.

Algorithm 3.3 (Control evaluation)

1. measure the state $x(t)$ at time instance t
2. search for the index set of regions $\mathcal{I}$ such that $x(t) \in \mathcal{P}_i$ for all $i \in \mathcal{I}$
    IF $\mathcal{I} = \emptyset$ **THEN** problem infeasible **STOP**
    IF $|\mathcal{I}| > 1$ **THEN** pick one element $i^* \in \mathcal{I}$
3. apply the control input $u(t) = \mu_{i^*}(x(t))$ to the system
4. wait for the new sampling time $t + 1$, goto (1.)

The second step in Algorithm 3.3 is also known as the *point-location problem* and we will get back to this issue in more detail in Chapter 11.

3.2.1 Time-varying Model and System

If one considers a time-varying system *and* time-varying prediction model of the form

$$x(\tau+1) = f(x(\tau), u(\tau), \tau+t), \quad \text{where} \quad \begin{bmatrix} x(\tau) \\ u(\tau) \end{bmatrix} \in \mathbb{X} \times \mathbb{U},$$

$\tau \in [0, T]$, $t \in \mathbb{Z}$, and a measurement $x_t =: x(0)$ of the system's state at sampling time t, i.e. $\tau = 0$, the (on-line) optimization procedure results is a time-varying optimal control input sequence

$$U_T^*(x_t, t) := \{u(0), u(1), \ldots, u(T-1)\}.$$

The equivalent closed-form time-varying state-feedback control law then takes the form $\mu(x_t, t, \tau)$, where $\tau \in \{0, \ldots, T-1\}$, $t \in \mathbb{Z}$, $x_t \in \mathbb{X}$. The optimal input to the model is then specified by

$$u(\tau) = \mu(x_t, t, \tau).$$

From this follows that the closed-form state feedback control law for the corresponding receding horizon control policy is still an explicit function of the absolute time t and system's state x_t at sampling time t, i.e.

$$u(t) = \mu_{\text{RH}}(x_t, t) := \mu(x_t, t, 0),$$

where $u(t)$ is applied to the 'real' system at time t.

Therefore, the optimal lookup table becomes time-varying and thus, in all its generality, might change with time t. Under the consideration of a time-invariant underlying 'real' system, the prediction model can be either time-varying *or* time-invariant in order to permit the computation of a closed-form static optimal control lookup table of the form (3.2'). This discrepancy between a time-varying prediction model and a time-invariant 'real' system can for example be utilized in the context of periodic systems. In the remainder of this work the focus will be on time-invariant 'real' systems.

3.2.2 Comments on the Closed-form Solution

The main theme of this work is to efficiently compute, analyze, and post-process closed-form state feedback controllers for the class of *piecewise affine systems* (Chapter 4). For this reason, some advantages and also disadvantages in this matter should be mentioned.

Advantages

High sampling rates. As mentioned above, one of the prime motivations of computing a closed-form solution $\mu_{RH}(\cdot)$ is in order to enable receding horizon control to be used for systems with high sampling rates. On-line one 'just' needs to search for the index set $\mathcal{I}$ of $\mu_{RH}(\cdot)$ and apply the corresponding control action to the system, cf. Algorithm 3.3. This is usually a very fast operation compared to the corresponding on-line optimization procedure and an efficient search algorithm is explained in Chapter 11.

Furthermore, in the on-line receding horizon control implementation (Algorithm 3.2) it is difficult to estimate upper bounds on the time needed for the optimization procedure to give a solution. For the point-location search, however, this can directly be guaranteed.

Cheap implementation. Once the closed-form optimal solution is computed, it can e.g. be implemented into a microprocessor and cheaply be replicated in mass production, whereas the on-line counterpart usually implies expensive and large computational infrastructure.

Analysis. The closed-form solution allows an overall better insight and understanding into the receding horizon control properties, such as for example closed-loop stability, feasibility for all time, and robustness, as well as optimality, reachability and safety verification for all feasible states. (Parts of this list are treated in the following chapters.) Moreover, it is possible to perform a proper visualization of the behavior or the system compared to 'simple' trajectory simulations for finite set of initial conditions.

Post-processing. Strongly interlinked with the analysis capabilities a simplification or complexity reduction of the overall closed-form solution can be performed; be it an approximation or reduction of the representation of the state feedback control law, without loosing properties such as closed-loop stability or feasibility for all time, cf. for example Chapter 10.

Other post-processing issues are the computation of a search tree to efficiently perform the aforementioned point-location problem and the 'robustification' of a nominal controller.

Disadvantages

Optimization complexity. For the most general setup it seems that obtaining a closed-form solution to receding horizon control is an extremely difficult problem, and although for interesting control problem classes solutions and solvers exist, the time needed to solve the overall parametric optimization problem grows (theoretically) in the *worst case* exponentially with its problem size, i.e. for example the prediction horizon, the number of constraints, as well as state and input dimension. This limits the applicability of the technique to 'smaller' problems.

However, most (if not all) control problems seem to be far less complex and the worst case exponential growth scenario is a *very* conservative upper bound on the overall complexity of the problem and does not seem to appear.

Solution complexity. As the optimization complexity grows, typically also the solution complexity, i.e. the number of defining regions N_P as well as the storage space needed for the closed-form solution, grows exponentially with its parameters. This in turn influences the on-line controller evaluation.

All the above mentioned aspects have lead to active research in this field and a flood of extensions to the basic idea of closed-form controller solutions are available [Bao05, Gri04, Md05, Gey05, Bor03, Tøn00, Lin03, BF03].

3.3 Choice of the Cost Function

Usually the optimization problem to be considered in Step (2.) of Algorithm 3.2 is of the form

$$J_T^*(x(0)) := \min_{U_T} J_T(x(0), U_T) \quad (3.3a)$$

$$\text{subj. to} \begin{cases} x(t+1) = f(x(t), u(t)) \\ x(T) \in \mathcal{X}^f, \end{cases} \quad (3.3b)$$

where

$$J_T(x(0), U_T) := \ell_T(x(T)) + \sum_{t=0}^{T-1} \ell(x(t), u(t), t) \quad (3.3c)$$

3.3 Choice of the Cost Function

is the *cost function* (also called *performance index*), $\ell(\cdot,\cdot,\cdot)$ is the *stage cost* function, $\ell_T(\cdot)$ is the *final penalty cost* function (also called *terminal cost* function), U_T is the *optimization variable* defined as the control action input sequence

$$U_T := \{u(t)\}_{t=0}^{T-1},$$

$T < \infty$ is the *prediction horizon*, and $\mathcal{X}^f$ is some *terminal target set* in $\mathbb{R}^{n_x}$. The optimizer of (3.3) is denoted by $U_T^*(x(0))$ and problem (3.3) implicitly defines the *set of feasible initial states* $\mathcal{X}_T \subseteq \mathbb{R}^{n_x}$, i.e. $x(0) \in \mathcal{X}_T$ for which $U_T^*(x(0))$ is nonempty.

If the cost function $J_T(\cdot,\cdot)$ is motivated by real scenarios, such as e.g. physical or financial principles, seemingly all (commonly used) forms for the stage cost and final penalty cost have their justification.

Most often, however, optimality is an artificial concept used to control a system in some 'good' way with the possibility to tune and to fulfill specific design specification.

Quadratic Cost Function

The cost function composed of *quadratic forms* (also called *quadratic cost function* or *quadratic performance index*), i.e.

$$\ell_T(x(T)) := x(T)'Px(T), \quad \text{and} \tag{3.4a}$$
$$\ell(x(t),u(t),t) := x(t)'Q(t)x(t) + u(t)'R(t)u(t), \tag{3.4b}$$

where P, $Q(t)$, and $R(t)$ are symmetric and positive semidefinite, is traditionally the primary choice in most optimal control problem formulations due to its connection to the 'energy' of the respective signals $x(t)$ and $u(t)$ as well as due to the decent scalability and predictable tunability of the optimization problem [LS95, Lev96, ZDG96].

Many nice properties of the above optimization problem (3.3) can be obtained using the quadratic cost function if the system and matrices $Q(t)$ and $R(t)$ have a particular structure, such as for example a simple constrained linear system and $Q(t)$ and $R(t)$ being symmetric and positive definite. Among others, the solution is unique, its similarity to the well known linear quadratic regulator, the smooth behavior of the controlled system close to the reference point, and strong activity of the controller if the system is far from the target point.

For this type of cost function, the closed-form solution $\mu(\cdot)$ introduced in Section 3.2, is well studied for different system classes, cf. for example [Bao05, Gri04, Md05, Gey05, Bor03, Tøn00, Lin03, BF03, Tor03].

Piecewise Linear Cost Function

Comparably, only very few authors considered a cost function based on linear vector norms, called (piecewise) *linear cost function* or *linear performance index*, i.e.

$$\ell_T(x(T)) := \|Px(T)\|_p, \quad \text{and} \tag{3.5a}$$

$$\ell(x(t), u(t), t) := \|Q(t)x(t)\|_p + \|R(t)u(t)\|_p, \tag{3.5b}$$

where $p \in \{1, \infty\}$ and P, $Q(t)$, and $R(t)$ may be tall or wide real matrices of suitable dimension[2]. Recall from Definition 1.39 that the vector 1- and ∞-norm [HJ85] is defined as

$$\|z\|_1 := \sum_{i=1}^{n_z} |z_i| \quad \text{and} \quad \|z\|_\infty := \max_{1 \le i \le n_z} |z_i|,$$

respectively, where here z_i denotes the i-th element of the vector $z \in \mathbb{R}^{n_z}$.

Even though optimal control problems with linear performance index are notoriously difficult to tune and, depending on the problem class, the resulting controller might encounter chattering or idle behavior in the actuator, they are of theoretical and practical interest. Compared to the quadratic cost function, typically the closed-loop system evolution results in a slower control action if the system is far from the reference point and becomes very active (e.g. jumping or dead-beat) as it gets closer.

These optimal control problems can most often be solved using linear programs. Probably the first to propose an LP based solution for optimal control problems with linear cost function were Zadeh and Whalen [ZW62] in 1962.

In the following, the non-exhaustive list of examples will indicate that, whenever one wants to minimize *'amounts'*, *worst case errors*, *(sum of) absolute errors*, *outliers*, or e.g. economically motivated signals over time the choice of linear vector norm based cost functions is to be preferred to the quadratic cost function. For instance in order to avoid outliers of a signal the choice of a quadratic cost function results in simple smoothening of the outliers but not in their avoidance.

- Receding horizon control to maximize the profit for material and information flow in supply chain networks such as multi-product batch plants [LL06, PLYG03]

- Cross directional receding horizon control of the basis weight and moisture content of paper in a pulp and paper machine [DDP99, SDRW01]

---

[2] Usually $Q(t)$ and $R(t)$ are chosen to be of full-column rank, which is not strictly necessary, as will be seen in Chapter 6.

- Receding horizon flow-rate control for a continuous countercurrent chromatography separation unit applied in fine chemical and biotechnology industry [Erd04]

- Minimization of the amount (or deviation of the amount) of chemicals used in batch processes during process operation

- Fuel consumption control of an interceptor rocket [Ben03]

- Impact-time control and missile guidance [JLT06]

- Control of a robot arm to minimize the maximum component of the velocity of its joints in order to avoid breakage of the joints due to outliers while tracking a desired end effector trajectory [TW01]

- Trajectory planning for controlled switching systems and non-holonomic robotic systems, such as car-like robotic path planning with minimal curvature or aircraft navigation [EM99]

- Hybrid control to minimize the highest temperature of an industrial vaporizer [RH05]

- Optimal portfolio management in fixed income investments

It should be stressed that the optimal control problems considered in this manuscript are solely dedicated to this type of cost function. The reason is twofold: first, the solution is of theoretical and practical interest (as indicated above) and at the same time not well understood, and second, for the rather general class of constrained piecewise affine systems (Chapter 4) controller solutions and analysis tools can efficiently be obtained, as will be shown in Part II and III. Moreover, the aforementioned issue of scalability and tunability of optimal control problems for linear hybrid systems, such as piecewise affine systems, is still unclear and difficult.

3.4 Stability and Feasibility in Receding Horizon Control

Blindly applying the solution of the constrained finite time optimal control problem (3.3) (without additionally modifying the underlying optimization problem) in a receding horizon manner to the system does *not* necessarily guarantee closed-loop *stability* or *feasibility* for all time [Löf03, MRRS00].

Definition 3.4 (Feasibility for all time). In accordance to Definition 1.51, a receding horizon control problem is called *feasible at time t* if there exists a controller at time t for the measured state $x(t)$, which satisfies the state and input constraints over the considered prediction horizon T. A receding horizon control problem is called *feasible for all time* if it is feasible for all $t \geq 0$. □

Löfberg [Löf03, Example 2.1] shows that depending on the choice of the parameters of the cost function, the receding horizon control strategy might drive even a simple linear unconstrained system with quadratic cost function unstable. Furthermore, he demonstrates in his example that the set in the (T, R)-parameter space, for which the closed-loop controlled system is stable, is a non-convex and disconnected set.

Therefore, tuning the 'free' design parameters T, P, Q, R, and $\mathcal{X}^f$ in the optimization problem (3.3) with (3.4) (or (3.5)) in order to guarantee closed-loop stability and feasibility for all time for the class of constrained piecewise affine systems, see Chapter 4, is far from being a simple task and one of the major themes throughout this work.

The well known survey paper [MRRS00] is an often used reference for this matter. However, it should be pointed out that the stability results rely on the continuity[3] of the value function $J_T^*(\cdot)$ of the optimization problem (3.3). Unfortunately this issue is mostly overlooked in the hybrid systems and receding horizon control literature where, depending on the system class and cost function chosen, discontinuity of $J_T^*(\cdot)$ naturally appears. A relaxation to the above is presented in [GSD05], where $J_T^*(\cdot)$ is 'only' assumed to be continuous for an ε-neighborhood around the equilibrium point.

An exception is the overview by Lazar et al. [LHWB06, Laz06] on asymptotic and exponential stability for *non-smooth systems* using receding horizon control. The result is stated here for completeness without proof.

Theorem 3.5 (Asymptotic/exponential stability in receding horizon control). Consider the optimization problem (3.3) with the nonlinear (possibly discontinuous) discrete-time system

$$x(t+1) = f(x(t), u(t)) \quad \text{with} \quad \begin{bmatrix} x(t) \\ u(t) \end{bmatrix} \in \mathbb{X} \times \mathbb{U}.$$

Furthermore, assume the existence of a local (possibly nonlinear and/or discontinuous) controller

$$u(t) = \mu_{\text{loc}}(x(t)) \quad \text{with} \quad x(t) \in \mathcal{X}^f,$$

---

[3] In fact, most results on stability in general, and in receding horizon control in particular, rely on (Lipschitz) continuity of the value function, the state feedback control law, and/or the state-update function itself.

3.4 Stability and Feasibility in Receding Horizon Control

where $\mu_{\text{loc}}(\mathbb{0}_{n_x}) = \mathbb{0}_{n_u}$ and $\mathbb{0}_{n_x} = f(\mathbb{0}_{n_x}, \mathbb{0}_{n_u})$; and assume there exist K-class functions $\underline{\alpha}(\,\cdot\,)$ and $\overline{\alpha}(\,\cdot\,)$, see Definition 1.44, and a neighborhood of the equilibrium point $\mathcal{N}(\mathbb{0}_{n_x}) \subseteq \mathcal{X}_T$ with

$$\ell(x, u) \geq \underline{\alpha}(\|x\|), \quad \forall\, x \in \mathcal{X}_T \text{ and } \forall u \in \mathbb{U},$$
$$\ell_T(x) \leq \overline{\alpha}(\|x\|), \quad \forall\, x \in \mathcal{N}(\mathbb{0}_{n_x}),$$

where $\ell(\mathbb{0}_{n_x}, \mathbb{0}_{n_u}) = 0$, $\ell_T(\mathbb{0}_{n_x}) = 0$. Suppose

- $\mathcal{X}^f \subseteq \mathbb{X}$,

- $\mathbb{0}_{n_x} \in \text{int}(\mathcal{X}^f)$,

- $\mathcal{X}^f$ is a positively invariant set for the closed-loop system $x(t+1) = f(x(t), \mu_{\text{loc}}(x(t)))$,

- $\mathcal{X}^f \subseteq \mathbb{X}^{\mu_{\text{loc}}}$, where $\mathbb{X}^{\mu_{\text{loc}}}$ is the input admissible set (see Definition 2.4),

and

$$\ell_T\left(f(x, \mu_{\text{loc}}(x))\right) - \ell_T(x) \leq -\ell(x, \mu_{\text{loc}}(x)), \quad \forall\, x \in \mathcal{X}^f.$$

Then for a fixed prediction horizon $T \geq 1$ the following holds:

(a) If the optimization problem (3.3) is feasible at time $t^\star \in \mathbb{N}_{\geq 0}$ for state $x_{t^\star} \in \mathbb{X}$, the problem (3.3) is feasible at time $t^\star + 1$ for $f(x_{t^\star}, u(0))$, where $x_{t^\star} = x(0)$ and $u(0)$ is the first element of $U_T^\star(x_{t^\star})$. Moreover, the optimization problem (3.3) is *feasible* for all $x \in \mathcal{X}_T$.

(b) When applying the receding horizon control scheme to the system in closed-loop, the equilibrium point $\mathbb{0}_{n_x}$ is *attractive* in $\mathcal{X}_T$ and $\mathcal{X}_T$ is positively invariant [GSD05].

(c) When applying the receding horizon control scheme to the system in closed-loop, the equilibrium point $\mathbb{0}_{n_x}$ is *asymptotically stable* in the Lyapunov sense in $\mathcal{X}_T$, while satisfying the state and input constraints.

(d) If $\underline{\alpha}(\|x\|) := \underline{a}\|x\|^\gamma$ and $\overline{\alpha}(\|x\|) := \overline{a}\|x\|^\gamma$ for some positive constants $\underline{a}, \overline{a}, \gamma > 0$ then, when applying the receding horizon control scheme to the system in closed-loop, the equilibrium point $\mathbb{0}_{n_x}$ is *locally exponentially stable* in the Lyapunov sense in $\mathcal{X}_T$, while satisfying the state and input constraints. ∎

Also here, it should be stressed that Theorem 3.5 permits both $J_T^*(\,\cdot\,)$ and $f(\,\cdot\,,\,\cdot\,)$ to be discontinuous for any $x \neq \mathbb{0}_{n_x}$ and only requires continuity of $J_T^*(\,\cdot\,)$

and $f(\cdot, \cdot)$ at the single point $\mathbb{0}_{n_x}$. Moreover, note that the assumption $\mathbb{0}_{n_x} \in \text{int}(\mathcal{X}^f)$ is not strictly necessary, but for the case that $\mathbb{0}_{n_x}$ is on the boundary of $\mathcal{X}^f$ one would need to modify the notion of stability for the closed-loop system.

Simply speaking, Theorem 3.5 implies for receding horizon control, in order to guarantee *a priori* closed-loop stability and feasibility for all time, that, once $\ell(\cdot, \cdot)$ and T is chosen, one has to find a positively invariant $\mathcal{X}^f$ and a control Lyapunov function $\ell_T(\cdot)$, which bounds the infinite time horizon cost from above in $\mathcal{X}^f$.

On the other hand, Theorem 3.5 is typically rather conservative and choosing the appropriate parameters is difficult. The closed-form solution to receding horizon control, mentioned in Section 3.2, permits the proof of stability and feasibility for all time (as well as instability and infeasibility) of designed systems *a posteriori*. This is partially treated in Part III.

Note that if the optimization problem involves state and/or input constraints as well as a short prediction horizon, the asymptotic stability of the closed-loop system may have no robustness, also called *zero robustness* [GMTT04]. This means that small perturbations can keep the closed-loop controlled system inside the feasible set $\mathcal{X}_T$ but simultaneously far away from the equilibrium point. However, this can not happen when receding horizon control is applied to an LTI system with convex state and/or input constraints. A necessary condition for zero robustness in receding horizon control is that the value function is discontinuous for some point in the interior of $\mathcal{X}_T$.

4
PIECEWISE AFFINE SYSTEMS

Polyhedral piecewise affine systems or *piecewise affine systems* (PWA) [Son81, HDB01] for short are defined by partitioning the extended state-input space into polyhedral regions and associating with each region a different affine state update equation.

Piecewise affine systems are (under very mild assumptions) equivalent to many other hybrid system classes [vS00, Kam01], such as mixed logical dynamical systems [BM99a], linear complementary systems [Hee99], and max-min-plus-scaling systems [DV01] to name a few and thus form a very general class of linear hybrid systems.

Moreover, piecewise affine systems present themselves to be a powerful class for identifying or approximating generic nonlineare systems via multiple linearizations at different operating points [Son81, FTMLM03, RBL04]. Although hybrid systems (and in particular PWA systems) are a special class of nonlinear systems, most of the nonlinear system and control theory does not apply because it requires certain smoothness assumptions. For the same reason one also cannot simply use linear control theory in some approximate manner to design controllers for PWA systems.

Consider the class of discrete-time, stabilizable, linear hybrid systems that can be described as constrained continuous[1] *piecewise affine* (PWA) systems of the following form

$$x(t+1) = f_{\text{PWA}}(x(t), u(t)) \quad (4.1\text{a})$$

$$:= A_d x(t) + B_d u(t) + a_d, \quad \text{if } \begin{bmatrix} x(t) \\ u(t) \end{bmatrix} \in \mathcal{D}_d, \quad (4.1\text{b})$$

where $t \geq 0$, $x \in \mathbb{R}^{n_x}$ is the state, $u \in \mathbb{R}^{n_u}$ is the control input, the domain $\mathcal{D} := \bigcup_{d=1}^{N_\mathcal{D}} \mathcal{D}_d$ of f_{PWA} is a non-empty compact set in $\mathbb{R}^{n_x+n_u}$ with $N_\mathcal{D} < \infty$ the number of system dynamics, and $\{\mathcal{D}_d\}_{d=1}^{N_\mathcal{D}}$ denotes a polyhedral partition of the domain $\mathcal{D}$, where $\mathcal{D}_d$ is full-dimensional, the closure of $\mathcal{D}_d$ is given by

$$\bar{\mathcal{D}}_d := \left\{ \begin{bmatrix} x \\ u \end{bmatrix} \in \mathbb{R}^{n_x+n_u} \mid D_d^x x + D_d^u u \leq D_d^0 \right\}, \quad \text{and} \quad \text{int}(\mathcal{D}_d) \cap \text{int}(\mathcal{D}_j) = \emptyset,$$

for all $d \neq j$.

Note that linear *state constraints* ($x \in \mathbb{X} \subseteq \mathbb{R}^{n_x}$) and *input constraints* ($u \in \mathbb{U} \subseteq \mathbb{R}^{n_u}$) of the general form $C^x x + C^u u \leq C^0$ are incorporated in the description of $\mathcal{D}_d$.

Remark 4.1 (Discontinuous PWA system). It should be remarked that most algorithms presented in the following apply also to general discontinuous PWA systems and give often reasonable results. However, few theoretical guarantees can then be given. □

With d the *d*-th *mode*, also called *d*-th *dynamic*, of the system is denoted. As the system evolves with time, we clearly have that d is a function of time t

$$d(t) = \left\{ d \mid \begin{bmatrix} x(t) \\ u(t) \end{bmatrix} \in \mathcal{D}_d, \, d \in \{1, \ldots, N_\mathcal{D}\} \right\}.$$

Henceforth, for brevity, as the system evolves with time, $d(t)$ will simply be denoted with d.

The following is a standing assumption throughout Part II of the manuscript.

Assumption 4.2 (Equilibrium at the origin). The origin in the extended state-input space is an equilibrium point of the PWA system (4.1), i.e.

$$\mathbb{0}_{n_x+n_u} \in \mathcal{D} \quad \text{and} \quad \mathbb{0}_{n_x} = f_{\text{PWA}}(\mathbb{0}_{n_x}, \mathbb{0}_{n_u}), \quad (4.2)$$

where $\mathbb{0}_{n_x} := [0 \, 0 \, \ldots \, 0]' \in \mathbb{R}^{n_x}$. □

---
[1] Here a PWA system defined over a disjoint domain $\mathcal{D}$ is called continuous if $f_{\text{PWA}}(\bullet)$ is continuous over connected subsets of the domain.

The above assumption does not limit the scope of this work. For simplicity, only the cost that penalizes the deviation of the state and control action from the origin (equilibrium point) in the extended state-input space is considered. However, all presented results also hold for any non-zero equilibrium point since such problems are easily translated to the *'steer-to-the-origin' problem* by a simple linear substitution of the variables.

Recall that in this manuscript the following definition of feasibility from Section 1.4 will be used.

Definition 1.51 (Feasibility). Consider the general parametric optimization problem

$$\check{J}^*(x) = \min_{\theta} \quad \check{J}(\theta, x) \tag{1.10a}$$

$$\text{subj. to} \quad \begin{bmatrix} \theta \\ x \end{bmatrix} \in \check{\mathcal{D}}, \tag{1.10b}$$

where $\check{J}$ is a real-valued function defined over the domain $\check{\mathcal{D}} \subseteq \mathbb{R}^{n_\theta + n_x}$, $x \in \mathbb{R}^{n_x}$ denotes a parameter vector (here mostly the state vector) and $\theta \in \mathbb{R}^{n_\theta}$ denotes the optimization variable (usually includes the input variable u).

The optimization problem (1.10a)–(1.10b) is called *feasible* if $\check{\mathcal{D}} \neq \emptyset$. The point $x \in \mathbb{R}^{n_x}$ (resp. $\theta \in \mathbb{R}^{n_\theta}$) is called *feasible* if and only if there exists a $\theta \in \mathbb{R}^{n_\theta}$ (resp. $\exists x \in \mathbb{R}^{n_x}$) such that $[\theta'\ x']' \in \check{\mathcal{D}}$. □

Please note that the terminology of *stability* of a system, i.e. stability of the origin (as it is considered in this work) only makes sense for feasible trajectories. Therefore, trajectories leaving a feasible set can not be considered as 'unstable'. The terminology of stability (in the classical sense) is not defined outside of a feasible set.

4.1 Obtaining a PWA Model

Depending on the structure of the physical hybrid system, it is often a challenging task to directly obtain a corresponding piecewise affine model of the system. A small number of software tools are available to automate the translation procedure from a 'natural' modeling language, which for example includes IF-THEN-ELSE structures and propositional logic, to the different aforementioned hybrid modeling classes. A list of modeling, analysis and synthesis tools for the class of hybrid systems is given e.g. in [HyS].

Here two software packages to fulfill this task should be mentioned. Both tools are freely available and are shipped as part of the Multi-Parametric Toolbox (MPT) [KGB04, KGBM03, KGBC06] for MATLAB®. (The reader is referred to the respective papers, manuals, and web-sites for further detail.)

The *HYbrid System DEscription Language* (HYSDEL) [Tor02, TB04, TBB⁺02] allows the user to describe the hybrid system in a high level intuitive textual form. The HYSDEL compiler then translates this form into a corresponding PWA or MLD model.

A different approach is taken in the *Hybrid Identification Toolbox* (HIT) [FT05] for MATLAB®, where clustering based algorithms reconstruct a PWA model from noisy input/output-data samples of the hybrid system.

Part II

Optimal Control of Constrained Piecewise Affine Systems

5
INTRODUCTION

In the last few years several different techniques have been developed for the analysis of and controller synthesis for hybrid systems [vS00, Hee99, Son81, LTS99, BZ00, BBM00, Bor03, Joh03]. A significant amount of the research in this field has focused on solving constrained optimal control problems, both for continuous-time and discrete-time hybrid systems.

As mentioned above, we here consider the class of discrete-time linear hybrid systems. In particular the class of constrained *piecewise affine* (PWA) systems that are obtained by partitioning the extended state-input space into polyhedral regions and associating with each region a different affine state update equation, cf. [Son81, HDB01]. As shown in [HDB01], the class of piecewise affine systems is of rather general nature and equivalent to many other hybrid system formalisms, such as for example mixed logical dynamical systems or linear complementary systems.

For piecewise affine systems the *constrained finite time optimal control* CFTOC problem can be solved by means of multi-parametric programming [Bor03]. The solution is a time-varying piecewise affine state feedback control law and can be computed by using *multi-parametric mixed-integer quadratic programming* (mp-MIQP) for a quadratic performance index and *multi-parametric mixed-integer linear programming* (mp-MILP) for a (piecewise) linear performance index, cf. [Bor03, DP00].

As recently shown by Borrelli *et al.* [BBBM03] for a quadratic performance index and in Chapter 6 [BCM03] a well as in [KM02] for a linear performance index, it is possible to obtain the optimal solution to the CFTOC problem without the use of integer programming. The authors in [BBBM03] and in Chapter 6 ([BCM03]) propose efficient algorithms based on a dynamic programming strategy combined with *multi-parametric quadratic* or *linear program* (mp-QP or mp-LP) solvers.

Unfortunately, stability and feasibility (constraint satisfaction) for all time of the closed-loop system are not guaranteed if the solution to the CFTOC problem is used without care in a *receding horizon control* (RHC) strategy [Bor03, Löf03]. To remedy this deficiency various schemes have been proposed in the literature. For constrained linear systems stability can be (artificially) enforced by introducing 'proper' terminal set constraints and/or a terminal cost to the formulation of the CFTOC problem [MRRS00]. For the class of constrained PWA systems very few and restrictive stability criteria are known, e.g. [BM99a, MRRS00]. Only recently ideas used for enforcing closed-loop stability of the CFTOC problem for constrained linear systems have been extended to PWA systems [GKBM04]. Unfortunately the technique presented in [GKBM04] introduces a certain level of suboptimality in the solution.

The main advantages of the infinite time solution, compared to the corresponding finite time solution of the optimal control problem, are the inherent guaranteed stability and feasibility for all time as well as optimality of the closed-loop system [SD87, MRRS00, BGW90, GBTM04]. However, *in general*, obtaining the solution is a very difficult task and the solution is most often too complex for a real implementation.

This part of the manuscript presents novel, computationally efficient algorithms to solve the *constrained finite time optimal control* problem (Chapter 6) and the *constrained infinite time optimal control* (CITOC) problem (Chapter 7) with a linear performance index for the class of constrained PWA systems. The algorithms combine a dynamic programming exploration strategy with a multi-parametric linear programming solver and basic polyhedral manipulation. In the case of the CITOC problem the developed algorithm guarantees convergence of the solution to the Bellman equation (if a bounded solution exists) which corresponds to the solution to the CITOC problem and thus avoids potential pitfalls of other conservative approaches. The algorithm cannot obtain optimal solutions that have an unbounded cost, but this is hardly a practical limitation since in most applications one wants to steer the state to some equilibrium point by spending a finite amount of 'energy'.

Remark 5.1 (Computational efficiency). The problems considered in this part of the manuscript belong to the class of (worst case) combinatorial problems, which in general have an exponential worst case complexity. The algorithms introduced for solving CFTOC and CITOC problems are 'efficient' in the sense that they are outperforming the mp-MILP formulation, which currently is the only other viable method of obtaining the closed-form solution to the problems at hand. Furthermore, unlike the mp-MILP approach, the implemented algorithms proposed here can cope with moderate size problems, cf. Section 5.1. □

5.1 Software Implementation

The presented algorithms for solving the CFTOC problem in Chapter 6 and the CITOC problem in Chapter 7 are implemented in the Multi-Parametric Toolbox (MPT) [KGB04, KGBC06] for MATLAB®. The toolbox can be downloaded free of charge at

> http://control.ee.ethz.ch/~mpt/

6

Constrained Finite Time Optimal Control

In this chapter a modification of the algorithm described in [BBBM03, Bor02] for computing the solution to the constrained finite time optimal control problem for discrete-time linear hybrid systems is presented. As opposed to the quadratic performance index used in the original algorithm here a performance index based on vector 1- and ∞-norms is used. The algorithm combines a dynamic programming strategy with a multi-parametric linear program solver. By comparison with results in the literature it is shown that the algorithm presented solves the considered class of problems in a computationally more efficient manner.

Consider the piecewise affine system[1] (4.1) and define the *constrained finite time optimal control* (CFTOC) problem

$$J_T^*(x(0)) := \min_{U_T} J_T(x(0), U_T) \tag{6.1a}$$

$$\text{subj. to } \begin{cases} x(t+1) = f_{\text{PWA}}(x(t), u(t)) \\ x(T) \in \mathcal{X}^f, \end{cases} \tag{6.1b}$$

---

[1] Recall that the constraints on $x(t)$ and $u(t)$ are included in the description of the domain $\mathcal{D}$ of the state-update function $f_{\text{PWA}}(\bullet, \bullet)$, cf. page 40.

where

$$J_T(x(0), U_T) := \ell_T(x(T)) + \sum_{t=0}^{T-1} \ell(x(t), u(t)) \tag{6.1c}$$

is the *cost function* (also called *performance index*), $\ell(\cdot, \cdot)$ is the *stage cost* function, $\ell_T(\cdot)$ is the *final penalty cost* function (also called *terminal cost* function), U_T is the *optimization variable* defined as the input sequence

$$U_T := \{u(t)\}_{t=0}^{T-1}, \tag{6.2}$$

$T < \infty$ is the *prediction horizon*, and $\mathcal{X}^f$ is a compact *terminal target set* in $\mathbb{R}^{n_x}$. The optimal value of the cost function, denoted with $J_T^*(\cdot)$, is called the *value function*. The optimization variable that achieves $J_T^*(\cdot)$ is called the *optimizer* and is denoted with

$$U_T^*(x(0)) := \{u^*(t)\}_{t=0}^{T-1}.$$

With a slight abuse of notation, when the CFTOC problem (6.1a)–(6.1c) has multiple solutions, i.e. when the optimizer is not unique, $U_T^*(x(0))$ denotes one (arbitrarily chosen) realization from the set of possible optimizers.

In the rest of this part of the manuscript the focus lies on the following restriction to the very general CFTOC problem (6.1a)–(6.1c)

Problem 6.1 (PWA system, 1-/∞-norm based cost). Consider the CFTOC problem (6.1a)–(6.1c) with

$$\ell(x(t), u(t)) := \|Qx(t)\|_p + \|Ru(t)\|_p \quad \text{and} \tag{6.3a}$$

$$\ell_T(x(T)) := \|Px(T)\|_p, \tag{6.3b}$$

where $\|\cdot\|_p$ with $p \in \{1, \infty\}$ denotes the standard vector 1- or ∞-norm [HJ85]. □

Note that it is common practice to use the term *linear performance index* when referring to (6.1c) using (6.3) even though, strictly speaking, the cost function $J_T(x(0), U_T)$ is a piecewise affine or piecewise linear function of its arguments.

Problem 6.1 implicitly defines the set of feasible initial states $\mathcal{X}_T \subseteq \mathbb{R}^{n_x}$ ($x(0) \in \mathcal{X}_T$) and the set of feasible inputs $\mathcal{U}_{T-t} \subseteq \mathbb{R}^{n_u}$ ($u(t) \in \mathcal{U}_{T-t}$, $t = 0, \ldots, T-1$). Our goal is to find an explicit (closed-form) expression for the set $\mathcal{X}_T$, and for the functions $J_T^* : \mathcal{X}_T \to \mathbb{R}$ and $u^*(t) : \mathcal{X}_T \to \mathcal{U}_{T-t}$, $t = 0, \ldots, T-1$.

Remark 6.2 (Infimum problem). Strictly speaking one should formulate Problem 6.1 as a search for the infimum rather than for the minimum, however, in this case the cost function $J_T(\cdot, \cdot)$ comprises a finite number of linear vector norms. Furthermore, since the set $\mathcal{X}^f$ is compact, $f_{\text{PWA}}(\cdot, \cdot)$ is continuous over connected subsets of $\mathcal{D}$, $N_{\mathcal{D}}$ is finite, and the sets $\mathcal{D}_d$, $d = 1, \ldots, N_{\mathcal{D}}$, are compact, it follows that the feasible space is compact. Consequently, the Weierstrass existence theorem [Ber95] guarantees that the minimum and infimum problem are equivalent, i.e. the infimum is obtained for some finite $U_T^*(x(0))$. □

Remark 6.3 (Choice of P, Q, R). The CFTOC Problem 6.1 can be posed and solved for any choice of the matrices P, Q, and R. However, from a practical point of view if we want to avoid unnecessary controller action while steering the state to the origin, the choice of a full column rank R is a necessity. Moreover, for stability reasons (as it will be shown in Chapter 7) a full column rank Q is assumed. □

Remark 6.4 (Time-varying system and/or cost). The CFTOC Problem 6.1 naturally extends to PWA systems and/or cost functions with time-varying parameters, i.e. $A_d(t)$, $B_d(t)$, $a_d(t)$, $\mathcal{D}_d(t)$, as well as $Q(t)$ and $R(t)$ for $t = 0, \ldots, T-1$. For simplicity the focus lies on the time-invariant case but the CFTOC problem with time-varying parameters is of the same form and complexity as the CFTOC problem with time-invariant parameters and therefore it can be solved in an analog manner. □

We summarize the main result concerning the solution to the CFTOC Problem 6.1 which has been proven in e.g. [May01a, Bor03].

Theorem 6.5 (Solution to CFTOC). A solution to the optimal control Problem 6.1 is a piecewise affine value function

$$J_T^*(x(0)) = \Phi_{T,i} x(0) + \Gamma_{T,i}, \quad \text{if} \quad x(0) \in \mathcal{P}_{T,i} \quad (6.4)$$

of the initial state $x(0)$ and the optimal input $u^*(t)$ is a closed-form time-varying piecewise affine function of the initial state $x(0)$

$$u^*(t) = \tilde{\mu}_T^*(x(0), t) = K_{T-t,i} x(0) + L_{T-t,i}, \quad \text{if} \quad x(0) \in \mathcal{P}_{T,i}, \quad (6.5)$$

where $t = 0, \ldots, T-1$ and $\{\mathcal{P}_{T,i}\}_{i=1}^{N_T}$ is a polyhedral partition of the set of feasible states $x(0)$

$$\mathcal{X}_T = \cup_{i=1}^{N_T} \mathcal{P}_{T,i}, \quad (6.6)$$

with the closure of $\mathcal{P}_{T,i}$ given by $\bar{\mathcal{P}}_{T,i} = \{x \in \mathbb{R}^{n_x} \mid P_{T,i}^x x \leq P_{T,i}^0\}$. ∎

As mentioned in Chapter 3, here with $\mu(\cdot)$ a generic closed-form *state feedback control law* is denoted that maps a set of (feasible) states $\mathcal{X} \subseteq \mathbb{X}$ to a set of (admissible) control actions $\mathcal{U} \subseteq \mathbb{U}$. Thus μ specifies the *control action* (or *input action*)

$$u(t) = \mu(x(t), t)$$

that will be chosen at time t when the state is $x(t)$.

For the simple case of a *constrained discrete-time LTI system* the solution to the CFTOC Problem 6.1 has the following properties

Theorem 6.6 (Properties of the CFTOC solution for LTI systems). Consider the optimal control Problem 6.1 with restriction $N_\mathcal{D} = 1$ and

$$x(t+1) = Ax(t) + Bu(t) \quad \text{if} \quad \begin{bmatrix} x(t) \\ u(t) \end{bmatrix} \in \mathcal{D}.$$

Then the set of feasible states $\mathcal{X}_T$ is a closed polyhedron in $\mathbb{R}^{n_x}$. The value function $J_T^*(\cdot)$ is convex and polyhedral piecewise affine over $\mathcal{X}_T$.

Moreover, if the optimal control function $\widetilde{\mu}_T^*(\cdot, t) : \mathcal{X}_T \to 2^{\mathbb{R}^{n_u}}$ for $t = 0, \ldots, T-1$ is a unique single-valued function $\widetilde{\mu}_T^*(\cdot, t) : \mathcal{X}_T \to \mathbb{R}^{n_u}$ for all $x(0) \in \mathcal{X}_T$ then $\widetilde{\mu}_T^*(x(0), t)$ is a continuous and polyhedral piecewise affine function of $x(0)$. Otherwise it is always possible to define an optimal control function $\widetilde{\mu}_T^*(x(0), t)$, which is a continuous and polyhedral piecewise affine function of $x(0)$ for all $x(0) \in \mathcal{X}_T$ and $t = 0, \ldots, T-1$. □

Proof. For the case of $N_\mathcal{D} = 1$ the CFTOC Problem 6.1 can be posed as a single mp-LP for which the optimizer and optimal solution has the aforementioned properties. Refer to Section 1.5 and [Bor03, Bao05, Gal95]. ∎

A discontinuous optimizer might lead to effects such as chattering or jumping in the control action. These issues can be remedied using an mp-LP algorithm based on lexicographic perturbation as proposed in [Jon05, JKM05]. Thus a unique, continuous, and piecewise affine control law $u^*(t) = \widetilde{\mu}_T^*(x(0), t)$ for all feasible $x(0) \in \mathcal{X}_T$ and $t = 1, \ldots, T-1$ can always be obtained.

6.1 CFTOC Solution via mp-MILP

One way of solving the constrained finite time optimal control problem (6.1a)–(6.1c) is by reformulating the PWA system (4.1) into a set of inequalities with integer variables ζ which act as switches between the different 'dynamics' $\mathcal{D}_d$ of the hybrid system. An appropriate modeling framework for such a class of systems is *mixed logical dynamical* (MLD) systems [BM99a], refer also to Chapter 4, where the switching behavior as well as the constraints of the system are modeled with inequality conditions. As mentioned above, Heemels *et al.* [HDB01] show the equivalence between PWA and MLD systems.

Using an MLD representation, the CFTOC Problem 6.1 can be stated in the following form

$$J_T^*(x(0)) := \min_{U_l} \; \|Px(T)\|_p + \sum_{t=0}^{T-1} \|Qx(t)\|_p + \|Ru(t)\|_p, \tag{6.7a}$$

$$\text{subj. to} \begin{cases} x(t+1) = \widetilde{A}x(t) + B^u u(t) + B^\zeta \zeta(t) + B^z z(t), \\ M^\zeta \zeta(t) + M^z z(t) \le M^u u(t) + M^x x(t) + M^0, \\ x(T) \in \mathcal{X}^f, \end{cases} \tag{6.7b}$$

where $p \in \{1, \infty\}$, $\zeta \in \{0,1\}^{n_\zeta}$ is a vector of integer variables, and $z \in \mathbb{R}^{n_z}$ represents a vector of auxiliary variables, cf. [BM99a].

By using an upper bound θ_t^x for each of the components, e.g. $\|Qx(t)\|_p \leq \theta_t^x$, of the cost function in (6.7a) and

$$x(t+1) = (\tilde{A})^t x(0) \sum_{j=0}^{t-1} (\tilde{A})^j \Big\{ B^u u(t-1-j) + B^\zeta \zeta(t-1-j) + B^z z(t-1-j) \Big\} \quad (6.8)$$

the CFTOC problem can be rewritten as a *mixed-integer linear program* (MILP) of the form

$$J_T^*(x(0)) = \min_\theta \quad c'\theta \quad (6.9a)$$

$$\text{subj. to} \quad \tilde{M}^\theta \theta \leq \tilde{M}^0 + \tilde{M}^x x(0), \quad (6.9b)$$

where $\tilde{M}^\theta$, $\tilde{M}^0$, and $\tilde{M}^x$ are matrices of suitable dimension containing the whole information on the state and input constraints, the weighting matrices P, Q, and R, as well as the state update equation (6.8) for the whole time horizon T. (Note that for the construction of $\tilde{M}^x$ it is necessary to compute $(\tilde{A})^t$ for all $t \in \{0, \ldots, T\}$.) Moreover, $c = [0'_{(n_u + n_\zeta + n_z) \cdot T}\ 1\ \ldots\ 1]'$, the optimization variable is of the form

$$\theta := \Big[u(0)', \ldots, u(T-1)', \zeta(0)', \ldots, \zeta(T-1)',$$

$$z(0)', \ldots, z(T-1)', \theta_0^{x'}, \ldots, \theta_T^{x'}, \theta_0^{u'}, \ldots, \theta_{T-1}^{u'} \Big]',$$

and $z(t)$ denotes an auxiliary continuous variable.

For a given, fixed initial state $x(0)$ the MILP[2] (6.9) can be solved in order to obtain the optimizer $\theta^*(x(0))$, which in turn provides the optimal control sequence $U_T^*(x(0))$. If $x(0)$ is considered as a 'free' parameter of the MILP (6.9) one has to solve a *multi-parametric MILP* (mp-MILP) for exploring the whole feasible state space. Dua and Pistikopoulos [DP00] proposed to split the original mp-MILP problem into two subproblems: an mp-LP and an MILP. The solution is found by recursion between these two subproblems by first fixing the integer variable and solving an mp-LP for this situation in order to explore and partition the feasible space. The intermediate solution gives an upper bound on the optimal cost. Then a new integer variable is fixed, an mp-LP is solved, and in case of overlapping polyhedral regions of the state space the resulting cost is compared with the previous one. In order to limit the exploration of the state space, which grows exponentially with the time horizon and the number of possible switching sequences, a branch and bound technique is applied.

---

[2] An efficient MILP-solver implementation can e.g. be found in the ILOG CPLEX® [ILO] optimization package.

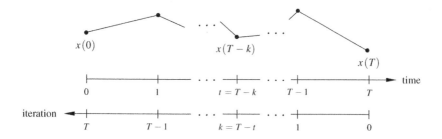

Fig. 6.1: Relation of the time axis and iteration step of the dynamic program.

6.2 CFTOC Solution via Dynamic Programming

Making use of Bellman's optimality principle [Bel57, BD62, Ber00], the CFTOC Problem 6.1 can be solved in a computationally efficient way by solving an equivalent *dynamic program* (DP) backwards in time [BCM03, Ber01, BS96]. The corresponding DP has the following form

$$J_k^*(x(t)) := \min_{u(t)} \quad \|Qx(t)\|_p + \|Ru(t)\|_p + J_{k-1}^*(f_{\text{PWA}}(x(t), u(t))), \quad (6.10a)$$

$$\text{subj. to} \quad f_{\text{PWA}}(x(t), u(t)) \in \mathcal{X}_{k-1} \quad (6.10b)$$

for $k = 1, \ldots, T$, with $t = T - k$, cf. Figure 6.1, and boundary conditions

$$\mathcal{X}_0 = \mathcal{X}^f \text{ and } J_0^*(x(T)) = \|Px(T)\|_p, \quad (6.10c)$$

where

$$\mathcal{X}_k := \{x \in \mathbb{R}^{n_x} \mid \exists u \in \mathbb{R}^{n_u}, f_{\text{PWA}}(x, u) \in \mathcal{X}_{k-1}\} \quad (6.10d)$$

is the set of all states at time $t = T - k$ for which the problem (6.10a)–(6.10c) is feasible.

Since $p \in \{1, \infty\}$ the dynamic programming problem (6.10) can be solved by multi-parametric linear programs, where the state $x(t)$ is treated as a parameter and the control input $u(t)$ as an optimization variable, cf. [Bor03, Bao05, BCM03]. By solving such programs at each iteration step k, going backwards in time starting from the target set $\mathcal{X}^f$, one obtains the set $\mathcal{X}_k \subset \mathbb{R}^{n_x}$, the optimal control law $u^*(t) : \mathcal{X}_k \to \mathcal{U}_k$, with $t = T - k$, and the value function $J_k^* : \mathcal{X}_k \to \mathbb{R}$ that represents the so called 'cost-to-go'. Properties of the solution are given in the following theorem, cf. e.g. [May01a, Bor03].

Theorem 6.7 (Solution to CFTOC via DP). The solution to the optimal control problem (6.10) with $p \in \{1, \infty\}$ is a piecewise affine value function

$$J_k^*(x(t)) = \Phi_{k,i} x(t) + \Gamma_{k,i}, \quad \text{if} \quad x(t) \in \mathcal{P}_{k,i} \quad (6.11)$$

and the optimal input $u^*(t)$ is a time-varying piecewise affine function of the state $x(t)$, i.e. it is given as a *closed-form state feedback control law*

$$u^*(t) = \mu_k^*(x(t)) = F_{k,i}x(t) + G_{k,i}, \quad \text{if} \quad x(t) \in \mathcal{P}_{k,i}, \tag{6.12}$$

where $k = 1, \ldots, T$ and $t = T - k$. $\{\mathcal{P}_{k,i}\}_{i=1}^{N_k}$ is a polyhedral partition of the set of feasible states $x(t)$ at time t

$$\mathcal{X}_k = \cup_{i=1}^{N_k} \mathcal{P}_{k,i}, \tag{6.13}$$

with the closure of $\mathcal{P}_{k,i}$ given by $\bar{\mathcal{P}}_{k,i} = \{x \in \mathbb{R}^{n_x} \mid P_{k,i}^x x \leq P_{k,i}^0\}$. ∎

Theorem 6.5 states that the solution to the CFTOC Problem 6.1, i.e. the optimal input sequence $U_T^*(x(0))$ given by (6.5), is a function of the initial state $x(0)$ only. On the other hand, Theorem 6.7 describes the solution to the dynamic program (6.10) as the optimal *state feedback control law* $\mu_k^*(x(t))$. Since we know that both solutions must be identical (assuming that the optimizer is unique), this implies that there is a connection between the matrices $K_{k,i}$ and $L_{k,i}$ in (6.5) and the matrices $F_{k,i}$ and $G_{k,i}$ in (6.12). It is easy to see that $K_{T,i} = F_{T,i}$ and $L_{T,i} = G_{T,i}$. To establish the connection for the other coefficients one would have to carry out the tedious sequence of substitutions $x(t) = f_{\text{PWA}}(x(t-1), \mu_{T-t+1}^*(x(t-1)))$, which would eventually express $x(t)$ in (6.12) as a function of $x(0)$ only. However, in this chapter the focus lies on the DP approach in solving the CFTOC problem and since both approaches give the same solution, we will not go beyond this note in establishing an explicit connection between those coefficients. Having this in mind, from this point onwards, when we speak of the solution to the CFTOC problem we consider the solution in the form given in Theorem 6.7.

6.3 An Efficient Algorithm for the CFTOC Solution

In order to present Algorithm 6.8 to solve the CFTOC problem via the dynamic program (6.10), some explanation of the notation and employed functions needs to be given.

In the first step ($k = 1$) of the DP (6.10), the cost-to-go function $J_0^*(\cdot)$ in (6.10c) is piecewise affine, the terminal region $\mathcal{X}^f$ is a polyhedron, and the constraints are piecewise affine. Thus the first DP iteration can be solved with $N_\mathcal{D}$ multi-parametric linear programs (mp-LPs), where $N_\mathcal{D}$ is the number of system dynamics. In the second DP iteration ($k = 2$) the cost-to-go function $J_1^*(\cdot)$ is polyhedral piecewise affine and the terminal set $\mathcal{X}_1$ is a union of N_1 polyhedra, where N_1 is the number of polyhedra over which $J_1^*(\cdot)$ is defined. Note that the constraints are still piecewise affine but $\mathcal{X}_1$ is not necessarily a convex

6.3 An Efficient Algorithm for the CFTOC Solution

set. The second DP can therefore be solved using $N_\mathcal{D} N_1$ mp-LPs. From $k = 3$ to T the 'nature' of the respective functions and underlying sets does not change.

In Algorithm 6.8, when SOLVE iteration k of a DP is mentioned, it is meant that several mp-LPs are formulated and a triplet of expressions for the value function, the optimizer, and the polyhedral partition of the feasible state space

$$\mathcal{S}_k^* := \left(J_k^*(\cdot),\ \mu_k^*(\cdot),\ \{\mathcal{P}_{k,i}\}_{i=1}^{N_k} \right)$$

is obtained, for example by using MPT [KGB04]. As indicated before, by inspection of the DP problem (6.10a)–(6.10c) it can be seen that at each iteration step k one is solving $N_\mathcal{D} N_{k-1}$ mp-LPs. After that, by using polyhedral manipulation one has to compare all generated *regions*[3], check if they intersect and remove the redundant ones, before storing a new partition that consists of N_k regions.

In the step where INTERSECT & COMPARE is performed, redundant polyhedra are removed, i.e. such polyhedra are removed that are completely covered with other polyhedra [Bao05, KGB04], which have a 'better' (meaning smaller) corresponding value function expression. If some polyhedron $\mathcal{P}_{k,i}$ is only partially covered with 'better' regions, the part of $\mathcal{P}_{k,i}$ with the smaller value function can be partitioned into a set of convex polyhedra. Thus the polyhedral nature of the feasible state space partition is preserved in each iteration.

Algorithm 6.8 (Generating the CFTOC solution)

INPUT $f_{\text{PWA}}(x,u),\ \{\mathcal{D}_d\}_{d=1}^{N_\mathcal{D}},\ p,\ P,\ Q,\ R,\ T,\ \mathcal{X}^f$
OUTPUT The CFTOC solutions $\mathcal{S}_T^*, \ldots, \mathcal{S}_1^*$

LET $\mathcal{S}_0^* \leftarrow \left(J_0^*(x) := \|Px\|_p,\ \mu_0^*(x) := \mathbb{0}_{n_u},\ \mathcal{X}_0 := \mathcal{X}^f \right)$
FOR $k = 1$ TO T
  FOR $d = 1$ TO $N_\mathcal{D}$
    FOR EACH $\mathcal{P}_{k-1,i} \in \mathcal{X}_{k-1}$

$$s_{d,i} \leftarrow \text{SOLVE } \min_u \|Qx\|_p + \|Ru\|_p + J_{k-1}^*(f_{\text{PWA}}(x,u)),$$
$$\text{subj. to } \begin{cases} [x'\ u']' \in \mathcal{D}_d, \\ f_{\text{PWA}}(x,u) \in \mathcal{P}_{k-1,i} \end{cases}$$

    END
  END
  LET $\mathcal{S}_k^* \leftarrow$ INTERSECT & COMPARE $\{s_{d,i}\}$
END

---

[3] With the notion of 'region' it is not only meant the state space partition but also the corresponding value function defined over it. Thus the comparison operation is also performed on the different value function expressions.

Just recently a modification to Algorithm 6.8 was described in [BGBM05]. The authors of [BGBM05] propose to shift the computationally expensive INTERSECT & COMPARE operation from each dynamic programming iteration step k to the very end of the algorithm and at the same time explore the convexity property of the involved mp-LPs. On one hand this approach eliminates a number of infeasible switching combinations and avoids the computationally expensive INTERSECT & COMPARE in intermediate iteration steps but at the same time it needs to perform the INTERSECT & COMPARE operation once in the worst case on $(N_\mathcal{D})^T$ state space partitions with each $N_{\mathcal{X}_i}$ regions ($i = 1,\ldots,(N_\mathcal{D})^T$), whereas Algorithm 6.8 basically performs the INTERSECT & COMPARE operation T times on a much smaller number of partitions. The success of one or the other approach is highly dependent on the number of dynamics $N_\mathcal{D}$, the prediction horizon T, and the geometrical structure of the problem in intermediate steps and thus it is impossible to rigorously favor one or the other algorithm.

Most probably a mixture of both ideas will lead to a computationally reliable and more efficient computation of the solution. This would mean a restart of the modified algorithm proposed in [BGBM05] after $\tilde{T}_k$ iteration steps, where $\sum_k \tilde{T}_k = T$, using the value function solution at the end of iteration $k - 1$, as an initialization of the value function for the next 'batch' of $\tilde{T}_k$ dynamic programs.

6.4 Comments on the Dynamic Programming Approach for the CFTOC Problem

In this section, some general remarks on important issues regarding the dynamic programming based technique, compared to the mp-MILP approach described in Section 6.1, will be given.

Intermediate Solutions

An important advantage of the dynamic programming approach, compared to the approach based on multi-parametric mixed-integer programming, shortly described in Section 6.1, is that after every iteration step, starting from $k = 1$ to $k = T$, the data of all the intermediate optimal control laws, the polyhedral partitions of the state space, and the piecewise affine value functions are available. Thus, after completing T dynamic programming iterations, the solutions to T different CFTOC problems with time horizons varying from 1 to T are simultaneously available and can be used for analysis and control purposes. Furthermore, one can in an intermediate step test if the solution to the $k^\star$-th DP step ($k^\star < T$) is already satisfactory in terms of, e.g., complexity and performance and thus can avoid further unnecessary computations.

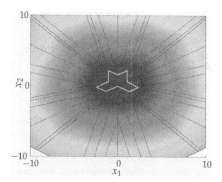 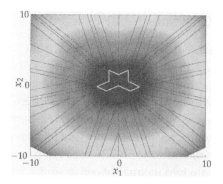

Fig. 6.2: State space partition of the finite time solution $J_T^*(x)$ for $T = 4$ for example (6.16) derived with the dynamic programming algorithm. Same color corresponds to the same cost value. The white marked region is identical to the infinite time solution.

Fig. 6.3: State space partition of the finite time solution $J_T^*(x)$ for $T = 5$ for example (6.16) derived with the dynamic programming algorithm. Same color corresponds to the same cost value. The white marked region is identical to the infinite time solution.

Infinite Time Horizon Solution

In addition, these intermediate solutions make it possible to detect if the solution for a specific time horizon is identical to the *infinite time solution* ($T \to \infty$), i.e. if for some $T = T_\infty$ the value function, as a function of the initial state $x(0)$ for all feasible states $x(0)$, is identical to the value function for $T = T_\infty + 1$.

In some parts of the state space, especially in the regions around the origin, it is likely that in two successive steps of the dynamic programming algorithm identical 'regions' – in terms of the regions' dimensions and the associated value function – are generated. Such a case is e.g. depicted in Figure 6.2 ($T = 4$) and Figure 6.3 ($T = 5$) for example (6.16) on page 60, where the white encircled regions are identical.

However, note that at some future iteration step the solution may change again and only when the piecewise affine value function converges on the whole feasible state space can one claim that the infinite time solution is obtained in any part of the state space. As a consequence it would be wrong to deduce that the infinite time solution was obtained in parts of the state space for some $T < T_\infty$. Such a claim can only be made a posteriori, i.e. after computing the solution to the CFTOC problem with $T \geq T_\infty$.

For further detail on the infinite time solution and explanation see Chapter 7. A modification of the algorithm in this chapter, that aims directly for the construction of the infinite time solution in a computationally more efficient manner by limiting the exploration of the state space in intermediate iteration steps of the dynamic programming algorithm is presented in Chapter 7.

Computational Efficiency

One of the main reasons for the efficiency of the dynamic programming algorithm compared to the mp-MILP based algorithm, is that the dynamic programming approach solves the CFTOC problem in the extended (x, u)-space with inequality constraints, whereas the mp-MILP based approach solves the same problem in the extended (x, u, ζ, z)-space including equality constraints, which are numerically much harder to treat and increase the size of the problem dramatically; this is because (usually) every equality constraint is translated into two inequality constraints for each time step for the whole time horizon.

Moreover, another numerical reliability issue of the dynamic programming based method is that for large time horizons T a computation of the possibly ill-conditioned matrices $(\widetilde{A})^t$ with $t = 1, \ldots, T$ in (6.8) is not needed while it is inherent in the MLD structure for the mp-MILP based method.

Thirdly, due to the possible degeneracy of linear programs it is impossible to provide a unique solution in *all* cases[4]. Therefore, it is very difficult to give a minimal representation for all possible occurring problems. However, the obtained number of polyhedral regions in the DP approach for any tested computed solution was always smaller than in the mp-MILP approach. This is due to the recursive structure of the mp-MILP method where unnecessary slicings of the state space are introduced. This fact is also reflected in the much larger memory demand of the mp-MILP approach as the time horizon increases, confer e.g. Table 6.1 on page 61.

6.5 Receding Horizon Control

In the case that the *receding horizon* (RH) control policy (described in Chapter 3) is used in closed-loop, the control is given as a time-invariant state feedback control law of the form

$$\mu_{\text{RH}}(x(t)) := \mu_T^*(x(t)) = F_{T,i} x(t) + G_{T,i}, \quad \text{if} \quad x(t) \in \mathcal{P}_{T,i} \quad (6.14)$$

with $u(t) = \mu_{\text{RH}}(x(t))$ and the time-invariant value function is

$$J_{\text{RH}}(x(t)) := J_T^*(x(t)) = \Phi_{T,i} x(t) + \Gamma_{T,i}, \quad \text{if} \quad x(t) \in \mathcal{P}_{T,i} \quad (6.15)$$

for $t \geq 0$, where $\{\mathcal{P}_{T,i}\}_{i=1}^{N_T}$ is a polyhedral partition of the set of feasible states $\mathcal{X}_T = \cup_{i=1}^{N_T} \mathcal{P}_{T,i}$. Thus only $N_{\text{RH}} := N_T$ (in the worst case different) control laws have to be stored.

---

[4] A possible way out of this dilemma is e.g. a technique based on lexicographic perturbation to always obtain a particular unique solution of an mp-LP [Jon05, JKM05].

6.6 Examples

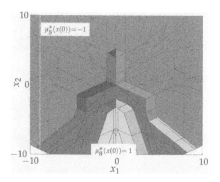

Fig. 6.4: State space partition of the finite time solution for $T = 8$ for example (6.16) derived with the dynamic programming algorithm. Same color corresponds to the same affine control law $\mu_8^*(x(0))$. The control law, comprising 19 different affine expressions, is defined over 262 polyhedral regions.

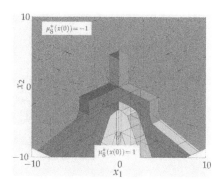

Fig. 6.5: State space partitioning of the finite time horizon solution for $T = 8$ for example (6.16) derived with the mp-MILP based algorithm. Same color corresponds to the same affine control law $\mu_8^*(x(0))$. The control law, comprising 19 different affine expressions, is defined over 2257 polyhedral regions.

Note that in general $J_{RH}(\cdot)$ in (6.15) does not represent the value function of the closed-loop system when the receding horizon control law $\mu_{RH}(\cdot)$ is applied because $J_{RH}(x(0))$ denotes the cost-to-go from $x(0)$ to $x(T)$ when the open-loop input sequence is applied. In the special case when the finite time solution is equivalent to the infinite time solution, i.e. $J_T^*(\cdot) \equiv J_\infty^*(\cdot)$ for some $T < \infty$, $J_{RH}(\cdot)$ in fact does represent the value function of the closed-loop system when applying $\mu_{RH}(\cdot)$, see also Remark 7.13.

Recall from Section 3.4 and Theorem 3.5, that if the receding horizon policy is applied to the system, a (possibly conservative) way to *a priori* guarantee closed-loop stability and feasibility for all time is to design the terminal target set $\mathcal{X}^f$ as an invariant set and the final penalty cost function $\|Px(T)\|_{\{1,\infty\}}$ as a control Lyapunov function over $\mathcal{X}^f$, which bounds the infinite time horizon cost from above.

Algorithm 8.12 in the Section 8.9 is a simple way to obtain such P an $\mathcal{X}^f$ for a class of piecewise affine systems. A more general approach for the here considered class of (general) piecewise affine systems would be to use the solution $\mathcal{C}_0$ of Algorithm 7.22 in Section 7.4 as PWA final penalty function $\ell_T(\cdot)$ and final target set $\mathcal{X}^f$.

6.6 Examples

In the following two examples using the aforementioned dynamic programming approach for solving the CFTOC problem are detailed. More examples are spread throughout the whole manuscript.

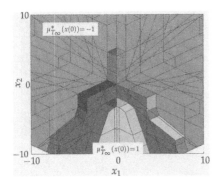 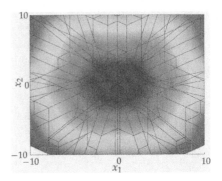

Fig. 6.6: State space partition of the infinite time solution ($T = T_\infty = 11$) for example (6.16) derived with the dynamic programming algorithm. Same color corresponds to the same affine control law $\mu^*_{T_\infty}(x(0))$.

Fig. 6.7: State space partition of the infinite time solution $J^*_{T_\infty}(x)$ with $T = T_\infty = 11$ for example (6.16) derived with the dynamic programming algorithm. Same color corresponds to the same cost value.

6.6.1 Constrained PWA Sine-Cosine System

Consider the piecewise affine system [BM99a]

$$\begin{cases} x(t+1) = \frac{4}{5}\begin{bmatrix} \cos\alpha(x(t)) & -\sin\alpha(x(t)) \\ \sin\alpha(x(t)) & \cos\alpha(x(t)) \end{bmatrix} x(t) + \begin{bmatrix} 0 \\ 1 \end{bmatrix} u(t), \\ \alpha(x(t)) = \begin{cases} \frac{\pi}{3} & \text{if } [1\ 0]x(t) \geq 0, \\ -\frac{\pi}{3} & \text{if } [1\ 0]x(t) < 0, \end{cases} \\ x(t) \in [-10,\ 10] \times [-10,\ 10], \quad \text{and} \quad u(t) \in [-1,\ 1]. \end{cases} \quad (6.16)$$

The CFTOC Problem 6.1 is solved with $Q = I_2$, $R = 1$, $P = 0_{2\times 2}$, and $\mathcal{X}^f = [-10,\ 10] \times [-10,\ 10]$ for $p = \infty$.

Figure 6.4 shows the state space partition of the finite time solution for $T = 8$ computed with the dynamic programming algorithm proposed in Section 6.3. The same color corresponds to the same affine control law $\mu^*_8(x(0))$. The control law, comprising 19 different affine expressions, is defined over 262 polyhedral regions. Each polyhedral region has a different affine value function expression assigned. Figure 6.5 shows the corresponding partition of the state space computed with the mp-MILP based algorithm presented in Section 6.1. Here the 19 different affine control laws were found in 2257 polyhedral regions. Unnecessary slicing of the state space was produced by the recursive structure of the mp-MILP algorithm.

In Table 6.1 the total number of polyhedral regions of the solution to the aforementioned problem for various horizons T obtained with the dynamic programming algorithm and the mp-MILP algorithm is reported. ⋆ denotes that the computation for the particular problem did not converge due to limited storage memory. Table 6.1 indicates the exponential growth of the number of regions using the mp-MILP approach of Section 6.1. Whereas the complexity

6.6 Examples

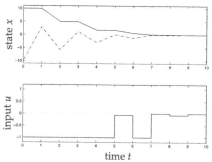

horizon T	# of regions for DP approach	# of regions for mp-MILP appr.
1	10	10
2	16	17
3	26	59
4	52	220
5	90	441
6	152	819
7	218	1441
8	262	2257
9	268	⋆
10	258	⋆
11	252	⋆

Fig. 6.8: State and control action evolution of the infinite time solution derived with the dynamic programming algorithm for example (6.16). Initial state $x(0) = [-10\ 10]'$.

Tab. 6.1: Comparison of the the number of regions for example (6.16). ⋆ denotes that the computation for the particular problem did not converge.

of the state space partition is significantly decreased using the DP approach of Section 6.3.

Figure 6.6 depicts the state space partition for the infinite time solution computed with the dynamic programming algorithm of Section 6.3. A posteriori it can be shown with the dynamic programming procedure that the finite time solution for any horizon $T \geq 11 = T_\infty$ is in fact identical to the infinite time solution of the constrained optimal control problem, cf. Section 6.4. The infinite time solution for this example was solved in 288.78 seconds on a Pentium M, 1.6 GHz (1GB RAM) machine running MATLAB® 7.0.4 SP2, the LP solver of NAG® [Num02], and MPT 2.6.1 [KGB04, KGBM03].

The same coloring scheme corresponds to the same affine control law. The optimal control law $\mu^*_{T_\infty}(x(0))$, comprising 23 different affine expressions, is defined over 252 polyhedral regions. Figure 6.7 reveals the corresponding value function for the state space partition. The same color corresponds to the same cost. (For a 3-dimensional visualization of the value function see Figure 7.5 on page 86.) The minimum cost is naturally achieved at the origin. Figure 6.8 shows the state and control action evolution with an initial state of $x(0) = [-10\ 10]'$ for the infinite time solution obtained with the dynamic programming procedure.

6.6.2 Car on a PWA Hill

Consider a car with mass m, motivated in [KGBC06], moving horizontally on a frictionless piecewise affine hill with different slopes α_i

$$\begin{aligned}\text{interval } \textcircled{1}: \quad & z \in [-\tfrac{1}{2},\ 1], \quad & \alpha_1 = 0°, \\ \text{interval } \textcircled{2}: \quad & z \in [-2,\ -\tfrac{1}{2}], \quad & \alpha_2 = 20°,\end{aligned} \qquad (6.17)$$

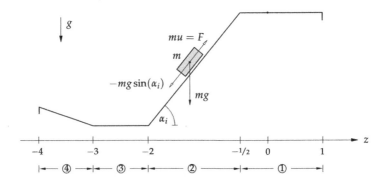

Fig. 6.9: Frictionless car moving on a piecewise affine hill.

interval ③ : $z \in [-3, -2]$, $\alpha_3 = 0°$,
interval ④ : $z \in [-4, -3]$, $\alpha_4 = -5°$,

cf. Figure 6.9. The continuous-time dynamics of the car are given by

$$\frac{dz(t)}{dt} = v(t), \quad \text{and} \quad m\frac{dv(t)}{dt} = F(t) - mg\sin(\alpha_{i(z(t))}),$$

where $z(t)$ denotes the horizontal position of the car at time t, $v(t)$ its horizontal velocity, g is the gravitational constant, and the slope angle α_i with $i = 1, \ldots, 4$ changes according to the car's position within the above mentioned intervals. The force $F(t)$ is assumed to be a linear function of the control action $u(t)$, i.e. $F(t) = m \cdot u(t)$, and is constrained by

$$|F(t)| \leq 2 \cdot m, \quad \text{and} \quad \left|\frac{dF(t)}{dt}\right| \leq 80 \cdot m.$$

The aim is to steer the car from any starting position $z(0) \in [-4, 1]$ and starting velocity $v(0) = 0$ to the origin on the top of the hill (and keeping it there) without 'dropping' from the piecewise affine 'environment', i.e. $z(t) \in [-4, 1]$ for all $t \geq 0$. Moreover, note that due to the constraints on the force $F(t)$ the car is not able to simply 'climb up' the steep 20°-slope of the second interval ($i = 2$).

For the here considered control purpose, the model is discretized with a sampling time of $T_s = 1/2$ and the following simple discontinuous piecewise affine model is obtained

$$x(t+1) = Ax(t) + Bu(t) + a_i, \tag{6.18a}$$

where $x(t) := [z(t), v(t)]'$ and

$$A = e^{\begin{bmatrix} 0 & T_s \\ 0 & 0 \end{bmatrix}} = \begin{bmatrix} 1 & 1/2 \\ 0 & 1 \end{bmatrix}, \quad B = \begin{bmatrix} T_s & \frac{1}{2}T_s^2 \\ 0 & T_s \end{bmatrix}\begin{bmatrix} 0 \\ 1 \end{bmatrix} = \begin{bmatrix} 1/8 \\ 1/2 \end{bmatrix}, \quad a_1 = a_3 = 0_2,$$

$$\tag{6.18b}$$

6.6 Examples

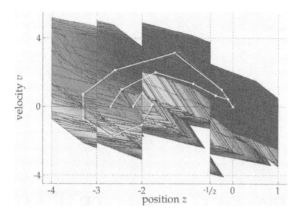

Fig. 6.10: State-space partition of the optimal closed-form receding horizon control solution $\mu_{RH}(\cdot)$ for the example in Section 6.6.2. The coloring corresponds to the open-loop value function $J_{RH}(\cdot)$. Trajectories indicate the stable closed-loop behavior.

$$a_{2,4} = \begin{bmatrix} T_s & \frac{1}{2}T_s^2 \\ 0 & T_s \end{bmatrix} \begin{bmatrix} 0 \\ -g\sin(\alpha_{2,4}) \end{bmatrix} = -gT_s \sin(\alpha_{2,4}) \begin{bmatrix} 1/2 T_s \\ 1 \end{bmatrix}. \quad (6.18c)$$

a_i with $i = 1, \ldots, 4$ is defined over the corresponding intervals as given in (6.17), $|u(t)| \leq 2$, and $|u(t+1) - u(t)| \leq 40$.

The CFTOC Problem 6.1 is solved for $p = 1$ and prediction horizon $T = 9$ with $Q = \text{diag}([100, 1]')$ and $R = 5$. The final penalty function $\|Px\|_1$ with $P \in \mathbb{R}^{2 \times 2}$ and $\mathcal{X}^f$ were computed as control Lyapunov function and positively invariant set, respectively, according to the simple Algorithm 8.12 in Section 8.9 in order to a priori guarantee closed-loop stability and feasibility for all time using receding horizon control, cf. also Theorem 3.5 and the paragraph below.

The state-space partition of the optimal closed-form receding horizon control solution $u(t) = \mu_{RH}(x(t))$ is depicted in Figure 6.10 and comprises 2083 polyhedral regions. The coloring corresponds to the open-loop value function $J_{RH}(\cdot)$ and ranges from dark blue to dark red, where dark blue refers to the value zero which is naturally obtained at the origin. Moreover, the coloring indicates the 'flow' of the closed-loop trajectories from dark red to dark blue. Sample trajectories show the stable closed-loop behavior.

As mentioned before, observe that simply 'climbing' the hill, starting at the bottom of the hill, is not possible. The optimal control strategy for initial states in the intervals 2–4 with zero initial velocity, drives the car to the left slope 4 in order to gain enough speed to cope with the steep slope 2.

7
Constrained Infinite Time Optimal Control

This chapter extends the ideas presented in Chapter 6 to the case of constrained infinite time optimal control problems for the class of discrete-time piecewise affine systems. The equivalence of the dynamic programming generated solution with the solution to the infinite time optimal control problem is shown. Furthermore, new convergence results of the dynamic programming strategy for general nonlinear systems and stability guarantees for the resulting possibly discontinuous closed-loop system are given. A computationally efficient algorithm to obtain the infinite time optimal solution, based on a dynamic programming exploration strategy with a multi-parametric linear programming solver and basic polyhedral manipulation, is presented.

7.1 Problem Formulation

As in the previous chapter, let us consider the piecewise affine system (4.1)

$$x(t+1) = f_{\text{PWA}}(x(t), u(t)) \qquad (4.1a)$$
$$= A_d x(t) + B_d u(t) + a_d, \quad \text{if } \begin{bmatrix} x(t) \\ u(t) \end{bmatrix} \in \mathcal{D}_d, \qquad (4.1b)$$

subject to state-input constraints

$$[x'\ u']' \in \mathcal{D} = \cup_{d=1}^{N_\mathcal{D}} \mathcal{D}_d \subset \mathbb{R}^{n_x+n_u}$$

7.1 Problem Formulation

as well as system dynamics $\{\mathcal{D}_d\}_{d=1}^{N_\mathcal{D}}$, cf. Chapter 4. By letting the time horizon $T \to \infty$ the cost function (6.1c) takes the following form (assuming that the limit exists)

$$J_\infty(x(0), U_\infty) := \lim_{T \to \infty} \sum_{t=0}^{T} \ell(x(t), u(t)), \tag{7.2}$$

where the function $\ell : \mathbb{R}^{n_x} \times \mathbb{R}^{n_u} \to \mathbb{R}_{\geq 0}$ is the *stage cost* and by $U_\infty := \{u(t)\}_{t=0}^{\infty}$ denotes the infinite *input/decision sequence*.

Moreover, (6.1a)–(6.1c) becomes the *constrained infinite time optimal control* (CITOC) problem

$$J_\infty^*(x(0)) := \min_{U_\infty} \quad J_\infty(x(0), U_\infty) \tag{7.3a}$$

$$\text{subj. to} \quad x(t+1) = f_{\text{PWA}}(x(t), u(t)), \tag{7.3b}$$

where $J_\infty^*(\cdot)$ is called the *(infinite time) value function* and $U_\infty^*(x(0)) := \{u^*(t)\}_{t=0}^{\infty}$ is the optimizer of (7.3).

For the restriction of a linear vector norm based stage cost one gets

Problem 7.1 (PWA system, 1-/∞-norm based cost). Consider the CITOC problem (7.3) with

$$\ell(x(t), u(t)) := \|Qx(t)\|_p + \|Ru(t)\|_p, \tag{7.4}$$

where $\|\cdot\|_p$ with $p \in \{1, \infty\}$ denotes the standard vector 1- or ∞-norm and Q is of full column rank. □

In order to guarantee closed-loop asymptotic stability, one needs to assume that Q is of full column rank as it will be shown in the following, cf. Lemma 7.4 and Remark 7.5. Additionally, in the infinite time case, it can be assumed that R is of full column rank even though both assumptions are not strictly needed, cf. Remark 6.3.

Assumption 7.2 (Boundedness). The CITOC problem (7.3), resp. Problem 7.1, is well defined, i.e. the minimum is achieved for some feasible input sequence $U_\infty^*(x(0))$, and $J_\infty^*(x(0)) < \infty$ for any feasible state $x(0)$ on a compact set $\mathcal{X}_\infty$. □

Without Assumption 7.2 the optimal control problem (7.3a)–(7.3b) would in general be undecidable, cf. [BT00, Sec. 4]. Additionally, Assumption 7.2 is reasonable from a practical point of view, since in most applications one wants to steer the state from some given state $x(0)$ or set to some equilibrium point (here the origin, cf. Assumption 4.2 and the paragraph below) by spending a finite amount of 'energy'. In the following example the reasoning behind Assumption 7.2 is additionally illustrated.

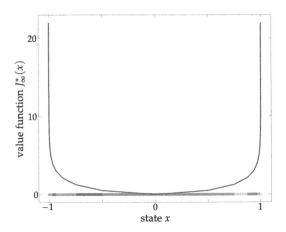

Fig. 7.1: δ-close approximation of the optimal value function $J^*_\infty(\cdot)$ for Example 7.3. The colored x-axis denotes the different regions over which the piecewise affine value function $J^*_\infty(\cdot)$ is defined.

Example 7.3 (Constrained LTI system). Consider the simple CITOC problem

$$J^*_\infty(x(0)) = \min_{U_\infty} \lim_{T \to \infty} \sum_{t=0}^{T} |x(t)|, \tag{7.5a}$$

$$\text{subj. to } \begin{cases} x(t+1) = 2x(t) + u(t), \\ |x(t)| \leq 1, \text{ and } |u(t)| \leq 1 \end{cases} \tag{7.5b}$$

for the constrained one-dimensional LTI system (7.5b).

Problem (7.5) is feasible for all initial states in $\bar{\mathcal{X}}_\infty = [-1, 1]$ and one can observe that the closed-loop system for the optimal state feedback control law

$$u^*(t) = \mu^*_\infty(x(t)) = \begin{cases} 1 & \text{if } x(t) \in \left[-1, -\frac{1}{2}\right], \\ -2x(t) & \text{if } x(t) \in \left[-\frac{1}{2}, \frac{1}{2}\right], \\ -1 & \text{if } x(t) \in \left[\frac{1}{2}, 1\right] \end{cases} \tag{7.6}$$

has three equilibria at $-1, 0,$ and 1. However, the closed-loop system is only asymptotically stable for the open set $\mathcal{X}_\infty = (-1, 1)$. Figure 7.1 illustrates that the optimal value function $J^*_\infty(x(0)) \to \infty$ as $x(0) \to \pm 1$ and therefore the problem is not well defined in the sense of Assumption 7.2. In practice, one can compute a δ-close approximation of $J^*_\infty(\cdot)$ on a closed subset $\mathcal{X}^\delta \subset \mathcal{X}_\infty$ as was done for obtaining Figure 7.1. Note that choosing any $R \neq 0$ does only influence the 'shape' of $J^*_\infty(\cdot)$ and $\mu^*_\infty(\cdot)$ but does not influence the above mentioned characteristic behavior of the solution.

7.2 Closed-loop Stability of the CITOC Solution

It should be remarked that most of the following results in Chapter 7 also hold (or are straight forwardly extended) for *general* discrete-time nonlinear systems

7.2 CLOSED-LOOP STABILITY OF THE CITOC SOLUTION

with *general* nonlinear stage costs and are not restricted to the considered class of PWA systems.

Lemma 7.4 (Stability of the CITOC solution). Consider the CITOC problem (7.3a)–(7.3b) and let its solution fulfill Assumption 7.2. Then the following holds:

(a) The origin $[x'\ u']' = \mathbb{0}_{n_x+n_u}$ is part of the infinite time solution, i.e. $\mathbb{0}_{n_x} \in \mathcal{X}_\infty$ with $J_\infty^*(\mathbb{0}_{n_x}) = 0$ and $u^*(t) = \mathbb{0}_{n_u}$ for $t \geq 0$.

(b) By applying the optimizer $U_\infty^*(x(0))$ to the system, any system state $x(0) \in \mathcal{X}_\infty$ is driven to the origin (*attractiveness*), i.e. if $x(0) \in \mathcal{X}_\infty$ then $x(t) \in \mathcal{X}_\infty$ for $t \geq 0$ and $\lim_{t\to\infty} x(t) = \mathbb{0}_{n_x}$.

(c) If $J_\infty^*(x)$ is continuous at the single point $x = \mathbb{0}_{n_x}$ (it can be discontinuous elsewhere), then by applying the optimizer $U_\infty^*(x(0))$ to the system for any $x(0) \in \mathcal{X}_\infty$, the equilibrium point $\mathbb{0}_{n_x} \in \mathcal{X}_\infty$ of the closed-loop system is *asymptotically stable* in the Lyapunov sense.

(d) If $J_\infty^*(x)$ is continuous at the single point $x = \mathbb{0}_{n_x}$ (it can be discontinuous elsewhere) and there exists a finite $\bar{\alpha} > 0$ such that $J_\infty^*(x) \leq \bar{\alpha}\|x\|_p$ for all $x \in \mathcal{X}_\infty$, then by applying the optimizer $U_\infty^*(x(0))$ to the system for any $x(0) \in \mathcal{X}_\infty$, the equilibrium point $\mathbb{0}_{n_x} \in \mathcal{X}_\infty$ of the closed-loop system is *exponentially stable* in the Lyapunov sense. □

Proof. (a) Because $[x'\ u']' = \mathbb{0}_{n_x+n_u}$ is an equilibrium point of the system (4.1) and $\ell(x,u) \geq 0$ for all $[x'\ u']' \in \mathcal{D}$, the minimum of $J_\infty(x(0), U_\infty) \geq 0$ for $x(0) = \mathbb{0}_{n_x}$ is achieved with e.g. $u^*(t) = \mathbb{0}_{n_u}$ for all $t \geq 0$. That means $J_\infty^*(\mathbb{0}_{n_x}) = 0$ which is the smallest possible value of $J_\infty(\cdot, \cdot)$.

(b) By Assumption 7.2 and $\ell(\cdot, \cdot) \geq 0$ we have $0 \leq J_\infty^*(\cdot) < \infty$, i.e. $J_\infty^*(\cdot)$ is bounded from above and below. Additionally, the sequence $\{J_T\}$ with $J_T := \sum_{t=0}^T \ell(x(t), u(t))$ for any sequence $\{(x(t), u(t))\}_{t=0}^T$, as T increases, is non-decreasing. Since we are using $U_\infty^*(x(0))$, the sequence $\{J_T\}$ converges to J_∞^*. Consequently, for every $\varepsilon_0 > 0$ there exists a $T_0 < \infty$ with $|J_T - J_\infty^*| < \varepsilon_0$ for all $T \geq T_0$. Therefore, necessarily $\lim_{T\to\infty} \ell(x(T), u(T)) = 0$ and because Q is of full column rank it follows $\lim_{T\to\infty} x(T) = \mathbb{0}_{n_x}$, cf. Definition 2.6.

(c) The origin $\mathbb{0}_{n_x} \in \mathcal{X}_\infty$ is an equilibrium point of the closed-loop system. Because Q is of full column rank, we have for any $t > 0$ and $x(0) \neq \mathbb{0}_{n_x}$ $(x(0) \in \mathcal{X}_\infty)$ that $J_\infty^*(x(t)) \leq J_\infty^*(x(0)) - \|Qx(0)\|_p < J_\infty^*(x(0))$. In general, for full column rank Q there exists a finite $\alpha_1 > 0$ such that $\alpha_1\|x\|_2 \leq \|Qx\|_p$. Therefore, we have $\alpha_1\|x(t)\|_2 \leq \|Qx(t)\|_p \leq J_\infty^*(x(t))$. On the other side, with Assumption 7.2 and the continuity of $J_\infty^*(x)$ at the single point $x = \mathbb{0}_{n_x}$ (it can be discontinuous elsewhere), there always exists a K-class function, cf. Definition 1.44, $\bar{J}(\cdot)$ with $J_\infty^*(x) \leq \bar{J}(\|x\|_2)$ for all $x \in \mathcal{X}_\infty$. Thus for all $t \geq 0$ and $x(0) \in \mathcal{X}_\infty$ it follows that $\|x(t)\|_2 \leq \frac{1}{\alpha_1}\bar{J}(\|x(0)\|_2)$. Clearly, for each $\bar{\varepsilon} > 0$ the choice of $\bar{\delta}(\bar{\varepsilon}) := \bar{J}^{-1}(\alpha_1 \cdot \bar{\varepsilon})$ satisfies the Lyapunov stability definition in Definition 2.7, i.e. for all $\bar{\varepsilon} > 0$ there exists $\bar{\delta}(\bar{\varepsilon}) > 0$ such that from $\|x(0)\|_2 < \bar{\delta}(\bar{\varepsilon})$ follows $\|x(t)\|_2 < \bar{\varepsilon}$ for all $t \geq 0$. Hence, the equilibrium $\mathbb{0}_{n_x}$ is a Lyapunov stable point and together with the attractiveness of $\mathbb{0}_{n_x}$ (Lemma 7.4(b)) we have that $\mathbb{0}_{n_x}$ is asymptotically stable.

(d) Together with the proof of Lemma 7.4(c) we get that $\underline{\alpha}\|x\|_p \leq \|Qx\|_p \leq J_\infty^*(x) \leq \bar{\alpha}\|x\|_p$ for some finite $\underline{\alpha} > 0$ and all $x \in \mathcal{X}_\infty$. (The existence of a finite $\underline{\alpha} > 0$ is always guaranteed.) Moreover, it follows that

$$J_\infty^*(f_{\text{PWA}}(x(t), U_\infty^*(x(t)))) - J_\infty^*(x(t)) = -\ell(x(t), u^*(t)) \leq -\|Qx(t)\|_p \leq -\underline{\alpha}\|x(t)\|_p$$

for all $x(t) \in \mathcal{X}_\infty$. Then Lemma 7.4(d) follows directly from Theorem 2.11 or [GSD05, Thm. 4.3.3]. ∎

Remark 7.5 (General nonlinear stage cost). Note that all stability and convergence results in Chapter 7 also hold for general nonlinear stage costs $\ell(\cdot, \cdot)$ for which a lower bounding K-class function $\underline{\alpha}(\cdot)$ exists, i.e.

$$\underline{\alpha}(\|x\|) \leq \ell(x, u) \quad \forall x \in \mathcal{X}, u \in \mathcal{U}.$$
□

Remark 7.6 (Continuity at $x = \mathbb{0}_{n_x}$). It should be pointed out that Lemma 7.4 only requires the continuity of $J_\infty^*(x)$ *at the single point* $x = \mathbb{0}_{n_x}$. No further (continuity) assumptions on $U_\infty^*(x)$ for closed-loop asymptotic stability are needed, which *is essential* for the considered problem class, where discontinuities of the solution may naturally appear. This is in contrast to most results in the literature where either a Lyapunov function and/or the closed-loop system description is assumed to be (Lipschitz) continuous for a neighborhood *around* $x = \mathbb{0}_{n_x}$, cf. e.g. [GSD05]. An exception in the literature, where only continuity at $x = \mathbb{0}_{n_x}$ is required, is [Laz06] which uses a different approach from the result presented in Lemma 7.4. □

The arguments in Remark 7.6 also hold for the later $J_{\infty,\text{DP}}^*(\cdot)$ and $\mu_{\infty,\text{DP}}^*(\cdot)$ of Lemma 7.7 in Section 7.3.

7.3 CITOC Solution via Dynamic Programming

Similar to the recasting of the CFTOC problem into a recursive dynamic program as presented in Section 6.2, it is possible to formulate for the CITOC problem (7.3) the corresponding *dynamic program* (DP) as follows

$$J_k(x(t)) := \min_{u(t)} \ell(x(t), u(t)) + J_{k-1}(f_{\text{PWA}}(x(t), u(t))) \tag{7.7a}$$

$$\text{subj. to} \quad f_{\text{PWA}}(x(t), u(t)) \in \mathcal{X}_{k-1}, \tag{7.7b}$$

7.3 CITOC Solution via Dynamic Programming

for $k = 1, \ldots, \infty$, with initial conditions

$$\mathcal{X}_0 = \{x \in \mathbb{R}^{n_x} \mid \exists u, [x'\ u']' \in \mathcal{D}\}, \text{ and} \tag{7.7c}$$

$$J_0(x) = 0 \quad \forall x \in \mathcal{X}_0. \tag{7.7d}$$

The set of all initial states for which the problem (7.7) is feasible at iteration step k is given by

$$\mathcal{X}_k := \{x \in \mathbb{R}^{n_x} \mid \exists u, f_{\text{PWA}}(x, u) \in \mathcal{X}_{k-1}\} \tag{7.8}$$
$$= \cup_{i=1}^{N_k} \mathcal{P}_{k,i}.$$

Let us define the feasible set of states as $k \to \infty$ by

$$\mathcal{X}_\infty := \lim_{k \to \infty} \mathcal{X}_k,$$

where here the limit is taken in the sense of the C- and Painlevé-Kuratowski convergence [LZ05]. Another more intuitive way to define the limit is through the Hausdorff distance [BBI01], i.e.

$$\lim_{k \to \infty} d_H(\mathcal{X}_k, \mathcal{X}_{k+1}) = 0,$$

with

$$d_H(\mathcal{Y}, \mathcal{Z}) := \max\{\sup_{y \in \mathcal{Y}} d(y, \mathcal{Z}), \sup_{z \in \mathcal{Z}} d(z, \mathcal{Y})\},$$

and

$$d(y, \mathcal{Z}) := \inf_{z \in \mathcal{Z}} \|y - z\|,$$

where here it needs to be assumed that all involved sets are compact. Furthermore, the limit value function of the dynamic program (7.7) is defined by

$$J^*_{\infty,\text{DP}} := \lim_{k \to \infty} J_k$$

and taken pointwise over $\mathcal{X}_\infty$. From $\mathbb{0}_{n_x} \in \mathcal{X}_0$, $J_0(\mathbb{0}_{n_x}) = 0$ and Assumption 4.2 it is easy to see that the following properties

$$\mathbb{0}_{n_x} \in \mathcal{X}_k \quad \text{and} \quad J_k(\mathbb{0}_{n_x}) = 0 \quad \text{for all} \quad k \geq 0 \tag{7.9}$$

of the dynamic program (7.7) hold.

Lemma 7.7 (Stability of the DP solution). Consider the DP problem (7.7). In addition, let $J^*_{\infty,\text{DP}}(x)$ be bounded on $\mathcal{X}_\infty$ and continuous at the single point $x = \mathbb{0}_{n_x}$. (It can be discontinuous otherwise.) Then, when applying the corresponding optimal control law $u^*(t) = \mu^*_{\infty,\text{DP}}(x(t))$ for all $x(t) \in \mathcal{X}_\infty$ to the system, the following holds:

(a) the limit value function $J^*_{\infty,\mathrm{DP}}(x)$ for all $x \in \mathcal{X}_\infty$, with $J^*_{\infty,\mathrm{DP}}(\mathbb{0}_{n_x}) = 0$, is a global Lyapunov function for the closed-loop system,

(b) any system state $x(0) \in \mathcal{X}_\infty$ is driven to the origin (*attractiveness*), i.e. if $x(0) \in \mathcal{X}_\infty$ then $x(t) \in \mathcal{X}_\infty$ for $t \geq 0$ and $\lim_{t \to \infty} x(t) = \mathbb{0}_{n_x}$, and

(c) the origin $x = \mathbb{0}_{n_x}$ is an *asymptotically stable* equilibrium point.

(d) Moreover, if there exists a finite $\bar{\alpha} > 0$ such that $J^*_{\infty,\mathrm{DP}}(x) \leq \bar{\alpha}\|x\|_p$ for all $x \in \mathcal{X}_\infty$, then the origin $x = \mathbb{0}_{n_x}$ is an *exponentially stable* equilibrium point. □

Proof. (a) From property (7.9) we have that $J^*_{\infty,\mathrm{DP}}(\mathbb{0}_{n_x}) = 0$. Because $J^*_{\infty,\mathrm{DP}}(\cdot)$ is a limit function of the dynamic program (7.7) we have

$$J^*_{\infty,\mathrm{DP}}(x) = \ell(x, \mu^*_{\infty,\mathrm{DP}}(x)) + J^*_{\infty,\mathrm{DP}}(f_{\mathrm{PWA}}(x, \mu^*_{\infty,\mathrm{DP}}(x)))$$

for all $x \in \mathcal{X}_\infty$. Thus, with $J^*_{\infty,\mathrm{DP}}(\cdot) \geq 0$, it follows that

$$-\Delta J^*_{\infty,\mathrm{DP}}(x) := J^*_{\infty,\mathrm{DP}}(x) - J^*_{\infty,\mathrm{DP}}(f_{\mathrm{PWA}}(x, \mu^*_{\infty,\mathrm{DP}}(x))) = \ell(x, \mu^*_{\infty,\mathrm{DP}}(x)) \geq \|Qx\|_p$$

for all $x \in \mathcal{X}_\infty$. Because Q is of full column rank, there exists a finite $\alpha_1 > 0$ with $\|Qx\|_p \geq \alpha_1 \|x\|_2$ and thus $-\Delta J^*_{\infty,\mathrm{DP}}(x) \geq \alpha_1 \|x\|_2$ for all $x \in \mathcal{X}_\infty$. This means that $-\Delta J^*_{\infty,\mathrm{DP}}(\cdot)$ is always bounded below by some K-class function. Similarly, we have that $J^*_{\infty,\mathrm{DP}}(x) \geq \ell(x, \mu^*_{\infty,\mathrm{DP}}(x)) \geq \|Qx\|_p \geq \alpha_2 \|x\|_2$ for all $x \in \mathcal{X}_\infty$ and some finite $\alpha_2 > 0$. By similar argument as in Lemma 7.4(c) there exists a K-class function $\bar{J}(\cdot)$ bounding $J^*_{\infty,\mathrm{DP}}(\cdot)$ from above. From these statements it follows directly that $J^*_{\infty,\mathrm{DP}}(\cdot)$ is a global Lyapunov function [Vid93, Laz06] for the closed-loop system.

(b)+(c) Because a global Lyapunov function exists (Lemma 7.7(a)) for the closed-loop system on the set $\mathcal{X}_\infty$ it follows immediately that the origin $\mathbb{0}_{n_x}$ is a global asymptotically stable [Vid93, Laz06], and thus if $x(0) \in \mathcal{X}_\infty$ then $x(t) \in \mathcal{X}_\infty$ for all $t \geq 0$ and $\lim_{t \to \infty} x(t) = \mathbb{0}_{n_x}$.

(d) With the proof of Lemma 7.7(a) we get that $\underline{\alpha}\|x\|_p \leq \|Qx\|_p \leq J^*_{\infty,\mathrm{DP}}(x) \leq \bar{\alpha}\|x\|_p$ and $\Delta J^*_{\infty,\mathrm{DP}}(x) \leq -\|Qx\|_p \leq -\underline{\alpha}\|x\|_p$ for all $x \in \mathcal{X}_\infty$ and some finite $\underline{\alpha} > 0$. (The existence of a finite $\underline{\alpha} > 0$ is always guaranteed.) Then Lemma 7.7(d) follows directly from Theorem 2.11 or [GSD05, Thm. 4.3.3]. ■

It should be remarked that Lemma 7.7(a) can be proven without imposing continuity of $J^*_{\infty,\mathrm{DP}}(x)$ at $x = \mathbb{0}_{n_x}$. The alternative proof is given in Appendix A. However, due to the analogy with Lemma 7.4(c) the above proof is presented here. Moreover, please confer Remark 7.6 about the continuity issue of the value function $J^*_{\infty,\mathrm{DP}}(\cdot)$, which seems to be most often overlooked in the (hybrid systems) control literature.

Note that in the infinite time case, in contrast to the finite time case discussed in Chapter 6, the equivalence of the solution of the dynamic program $J^*_{\infty,\mathrm{DP}}(\cdot)$ and the optimal solution $J^*_\infty(\cdot)$ of the CITOC problem is not immediate. Before

7.3 CITOC Solution via Dynamic Programming

this equivalence is proven, it is useful to introduce the following commonly used operators [Ber01, Ber00].

Definition 7.8 (Operator T and T_μ). For any function $J : \mathcal{X} \to \mathbb{R}_{\geq 0}$ lets define the following mapping

$$(TJ)(x) := \min_{u \in \mathcal{U}} \ell(x, u) + J(f_{\text{PWA}}(x, u)), \qquad (7.10)$$

where the set of feasible control actions $\mathcal{U} \subset \mathbb{R}^{n_u}$ is defined implicitly through the domains of $J(\cdot)$ and $f_{\text{PWA}}(\cdot, \cdot)$, i.e.

$$\mathcal{U} := \{u \in \mathbb{R}^{n_u} \mid \exists x, \ [x'\ u']' \in \mathcal{D},\ f_{\text{PWA}}(x, u) \in \mathcal{X}\}.$$

T transforms the function $J(\cdot)$ on $\mathcal{X}$ into the function $TJ : \tilde{\mathcal{X}} \to \mathbb{R}_{\geq 0}$ with some $\tilde{\mathcal{X}} \subseteq \mathbb{R}^{n_x}$. T^k denotes the k-times operator of T with itself, i.e.

$$(T^0 J)(\cdot) := J(\cdot) \quad \text{and} \quad (T^k J)(\cdot) := \left(T(T^{k-1}J)\right)(\cdot)$$

with $k \in \mathbb{N}_{\geq 0}$. Accordingly, lets use

$$(T_\mu J)(\cdot) := \ell(\cdot, \mu(\cdot)) + J(f_{\text{PWA}}(\cdot, \mu(\cdot)))$$

for any function $J : \mathcal{X} \to \mathbb{R}_{\geq 0}$ and any control function $\mu : \mathcal{X} \to \mathcal{U}$ defined on the state space $\mathcal{X}$. □

The DP procedure (7.7a)–(7.7d) can now be simply stated as

$$J_k(\cdot) := (TJ_{k-1})(\cdot) \quad \text{with} \quad J_0(\cdot) = 0, \qquad (7.7')$$

where $J^*_{\infty,\text{DP}} = \lim_{k \to \infty} T^k J_0$. The solution to the DP procedure, $J^*_{\infty,\text{DP}}(\cdot)$, satisfies the *Bellman equation*

$$J(\cdot) = (TJ)(\cdot), \qquad (7.11)$$

which is effectively being used as a stopping criterion to decide when the DP procedure has terminated.

Definition 7.9 (Stationarity). A control law $\mu(\cdot)$ is called *stationary* if for some function $J(\cdot)$ it follows $J(x) = (T_\mu J)(x)$ for all $x \in \mathcal{X}_\infty$. A function $J_\mu(\cdot)$ is called stationary if it fulfills the Bellman equation (7.11) for all $x \in \mathcal{X}_\infty$. □

To prove that the solution to the CITOC problem (7.3a)–(7.3b), $J^*_\infty(\cdot)$, is identical to the solution of the dynamic program (7.7a)–(7.7b), $J^*_{\infty,\text{DP}}(\cdot)$, one actually has to answer two important questions: first, under which conditions does the DP procedure (7.7a)–(7.7b) converge, and second, when is $J^*_{\infty,\text{DP}}(\cdot)$ a unique solution to the DP procedure (7.7a)–(7.7b).

Theorem 7.10 (Equivalence: CITOC solution–DP solution). Let the solution to the CITOC problem (7.3a)–(7.3b), $J_\infty^*(\cdot)$, satisfy Assumption 7.2 and let it be continuous *at* the single point $x = \mathbb{0}_{n_x}$. Let $J_{\infty,\mathrm{DP}}^*(\cdot)$ be the solution to the dynamic program (7.7a)–(7.7d). Then $J_\infty^*(x) = J_{\infty,\mathrm{DP}}^*(x)$ for all $x \in \mathcal{X}_\infty$. Moreover, the solution $J_{\infty,\mathrm{DP}}^*(\cdot)$ is a unique solution of the dynamic program (7.7a)–(7.7d). □

Proof. According to [Ber01, Sec. 3], the solution $J_\infty^*(\cdot)$ to the CITOC problem (7.3a)–(7.3b) satisfies the dynamic program (7.7a)–(7.7b), that is $J_\infty^*(\cdot) = (\mathbf{T} J_\infty^*)(\cdot)$. On the other hand, in general, the Bellman equation (7.11) may have no solution or it may have multiple solutions, but at most one solution has the property

$$\lim_{t \to \infty} J(x(t)) = 0, \qquad (7.12)$$

cf. [SL89, Sec. 4.]. Now, if the dynamic program (7.7a)–(7.7d) has a solution $J_{\infty,\mathrm{DP}}^*(\cdot) < \infty$ fulfilling property (7.12) then it is unique and according to [SL89, Thm. 4.3] this solution satisfies the CITOC problem. From Lemma 7.7 or Appendix A it follows immediately that the DP solution does in fact satisfy property (7.12). ■

Having established this result, in the following the solution to both problems, the CITOC problem (7.3a)–(7.3b) and the DP problem (7.7a)–(7.7d), will be denoted with $J_\infty^*(\cdot)$.

Lemma 7.11 (Optimal control law μ_∞^*). [Ber01, Prop. 3.1.3] A stationary control law $\mu_\infty^*(\cdot)$ is *optimal* if and only if $(\mathbf{T} J_\infty^*)(x) = (\mathbf{T}_{\mu_\infty^*} J_\infty^*)(x)$ for all $x \in \mathcal{X}_\infty$. In other words, $\mu_\infty^*(\cdot)$ is optimal if and only if the minimum of (7.7a) is obtained with $\mu_\infty^*(x)$ for all $x \in \mathcal{X}_\infty$. ■

As in the CFTOC case, the infinite time problem might have multiple optimizers $\mu_\infty^*(\cdot)$. With a slight abuse of notation, in the case when the optimizer is not unique, with $\mu_\infty^*(\cdot)$ one (arbitrarily chosen) realization from the set of possible optimizers is denoted.

Now we are ready to state the theorem that characterizes the optimal solution $J_\infty^*(\cdot)$ and the optimal state feedback control law $\mu_\infty^*(\cdot)$.

Theorem 7.12 (Solution to the CITOC and DP problem). Under Assumption 7.2 the solution to the infinite time optimal control Problem 7.1 is a piecewise affine value function

$$J_\infty^*(x(t)) = \Phi_{\infty,i} x(t) + \Gamma_{\infty,i}, \quad \text{if} \quad x(t) \in \mathcal{P}_{\infty,i} \qquad (7.13)$$

and the optimal state feedback control law is of the time-invariant piecewise affine form

$$u^*(t) = \mu_\infty^*(x(t)) = F_{\infty,i} x(t) + G_{\infty,i}, \quad \text{if} \quad x(t) \in \mathcal{P}_{\infty,i}, \qquad (7.14)$$

where $\{\mathcal{P}_{\infty,i}\}_{i=1}^{N_\infty}$ is a polyhedral partition of the set $\mathcal{X}_\infty$ of feasible states $x(t)$ at time $t \geq 0$. □

7.3 CITOC Solution via Dynamic Programming

Proof. Theorem 7.12 follows from the construction of the DP iterations (7.7a)–(7.7d), Theorem 6.7, Theorem 7.10, and Assumption 7.2. ∎

Note that N_∞ can tend to ∞, i.e. $N_\infty \to \infty$, even for very simple systems. However, especially for discrete-time systems which often have the property to be driven to the origin in a finite number of time steps, this does not necessarily have to be the case, cf. also the example Section 7.5. Furthermore, this does not limit the practicality of the proposed algorithms nor does it alter the properties or results in this chapter.

Remark 7.13 (Receding Horizon Control). Note that the infinite time optimal control law (7.14) has the same form as the corresponding *receding horizon* control policy (6.14). This means that with the optimal control law $\mu_\infty^*(\cdot)$ the closed-loop response using receding horizon control and open-loop response of the system (4.1) are identical. Moreover, the value $J_\infty^*(x(t))$ is the total 'cost-to-go' from $x(t)$ to the origin, applying the optimal control policy. □

7.3.1 Initial Value Function and Rate of Convergence

In the dynamic program (7.7a)–(7.7d), the value function iteration procedure is started with the zero-function (7.7d) as the initial condition. This is the 'classical', simplest, and most often used approach for initializing $J_0(\cdot)$ in the literature. Due to the monotonicity property of the operator $\mathbf{T}$, cf. [Ber01, Lem. 1.1.1], this guarantees the convergence ('from below') to the optimal solution $J_\infty^*(\cdot)$ of 'arbitrary' dynamic programming problems if $J_\infty(x) := \lim_{k\to\infty}(\mathbf{T}^k J_0)(x) < \infty$ is a stationary solution, i.e. $J_\infty(x) = (\mathbf{T}J_\infty)(x)$ for all feasible x, cf. [Ber01, Prop. 3.1.5]. However, in numerical computations it might happen that at some point of the DP iteration an erroneous $\widetilde{J}_k(\cdot)$, with $\widetilde{J}_k(x) > J_\infty^*(x)$ for some x, is computed and thus convergence ('from below') of the DP is not guaranteed. Moreover, in the case that a (tight) upper bound[1] $J_0(\cdot) \geq J_\infty^*(\cdot)$ is available, one might want to incorporate this information into the dynamic programming procedure because if convergence ('from above') can be guaranteed, due to the monotonicity property of the operator $\mathbf{T}$, it is likely that this 'speeds up' the convergence of the DP.

The following results show that one can in fact guarantee convergence to the optimal solution $J_\infty^*(\cdot)$ from (almost) arbitrary $J_0(\cdot)$; but first it is necessary to prove some preliminary result.

Lemma 7.14 (Lower bound). *Let the function* $\underline{J} : \mathcal{X}_\infty \to \mathbb{R}_{\geq 0}$ *with*

$$\underline{J}(0_{n_x}) = 0 \quad \text{and} \quad \underline{J}(x) \leq \ell(x,u) + \underline{J}(f_{\text{PWA}}(x,u)) \tag{7.15}$$

for all $(x,u) \in \mathcal{X}_\infty \times \mathcal{U}$ *then* $\underline{J}(x) \leq J_\infty^*(x)$ *for all* $x \in \mathcal{X}_\infty$. □

---
[1] A candidate for an upper bound is e.g. the cost corresponding to any stabilizing suboptimal control law.

Proof. By Lemma 7.11 the control input sequence $\{\mu_\infty^*(x(t))\}_{t=0}^\infty$ obtains the optimal value function $J_\infty^*(x(0))$ for all $x(0) \in \mathcal{X}_\infty$. Using a $\underline{J}(\cdot)$ with the properties given in Equation (7.15) it follows for the optimal sequence pair $\{(x(t), \mu_\infty^*(x(t)))\}_{t=0}^\infty$ that $\ell(x(t), \mu_\infty^*(x(t))) \geq \underline{J}(x(t)) - \underline{J}(x(t+1))$ for all $t \geq 0$. Moreover, under Assumption 7.2, from Lemma 7.4 we have $J_\infty^*(\mathbb{0}_{n_x}) = 0$ and $\lim_{t \to \infty} x(t) = 0$, i.e. $J_\infty^*(x(0)) < \infty$ for all $x(0) \in \mathcal{X}_\infty$. Therefore, it follows that

$$\infty > J_\infty^*(x(0)) = \lim_{T \to \infty} \sum_{t=0}^T \ell(x(t), \mu_\infty^*(x(t))) \geq \lim_{T \to \infty} \sum_{t=0}^T \underline{J}(x(t)) - \underline{J}(x(t+1))$$
$$= \lim_{T \to \infty} \underline{J}(x(0)) - \underline{J}(x(T)) = \underline{J}(x(0)) - \underline{J}(\mathbb{0}_{n_x}) = \underline{J}(x(0))$$

for all $x(0) \in \mathcal{X}_\infty$. ∎

Theorem 7.15 (Initial value function J_0). Let Assumption 7.2 hold, $J_\infty^*(x)$ be continuous at $x = \mathbb{0}_{n_x}$, and let $J_0 : \mathcal{X}_\infty \to \mathbb{R}_{\geq 0}$ with $J_0(\mathbb{0}_{n_x}) = 0$, $J_0(\cdot) < \infty$, and $\mathbb{0}_{n_x} \in \mathcal{X}_\infty$, be the initial value function of the dynamic program iteration (7.7a)–(7.7b). Moreover, let

$$J_\mu(x) := \lim_{k \to \infty} \left(\mathbf{T}^k J_0\right)(x) \tag{7.16}$$

be a finitely bounded limit function, i.e. $J_\mu(\cdot) < \infty$ with $J_\mu(\mathbb{0}_{n_x}) = 0$, and

$$\mu(x) := \arg\min_{u \in \mathcal{U}} \ell(x, u) + J_\mu(f_{\text{PWA}}(x, u)) \tag{7.17}$$

its corresponding stationary control law for all $x \in \mathcal{X}_\infty$. Then the following separate statements hold:

(a) If $J_0(\cdot)$ is some *(arbitrary) finitely bounded* function on $\mathcal{X}_\infty$, and if after some (possibly infinite) DP iterations $J_\mu(\cdot)$ is a stationary solution, i.e. $J_\mu(x) = (\mathbf{T} J_\mu)(x)$ for all $x \in \mathcal{X}_\infty$, then $J_\mu(x) = J_\infty^*(x)$ and $\mu(x)$ is an optimal control law for all $x \in \mathcal{X}_\infty$.

(b) If $J_0(\cdot)$ is some *(arbitrary) finitely bounded* function on $\mathcal{X}_\infty$ with either $0 \leq J_0(x) \leq J_\infty^*(x)$ for all $x \in \mathcal{X}_\infty$ or $J_0(x) \geq (\mathbf{T} J_0)(x)$ for all $x \in \mathcal{X}_\infty$ then the existence of $J_\mu(\cdot)$ is guaranteed, $J_\mu(x) = J_\infty^*(x)$, and $\mu(x)$ is an optimal control law for all $x \in \mathcal{X}_\infty$.

(c) If $J_0(\cdot)$ is a *finitely bounded realization* of some feasible control input sequence $\{\widetilde{\mu}(x(t))\}_{t=0}^\infty$ with $\widetilde{\mu}(x(t)) \in \mathcal{U}$, i.e.

$$J_0(x(0)) := \lim_{T \to \infty} \sum_{t=0}^T \ell(x(t), \widetilde{\mu}(x(t))) \quad \text{for all} \quad x(0) \in \mathcal{X}_\infty,$$

then the existence of $J_\mu(\cdot)$ is guaranteed, $J_\mu(x) = J_\infty^*(x)$, and $\mu(x)$ is an optimal control law for all $x \in \mathcal{X}_\infty$. □

7.3 CITOC Solution via Dynamic Programming

Proof.   (a) We have that $J_\mu(\cdot)$ is a stationary solution of the dynamic program (7.7a)–(7.7b), i.e. $J_\mu(x) = (\mathbf{T}J_\mu)(x)$ for all $x \in \mathcal{X}_\infty$ with $J_\mu(\mathbb{0}_{n_x}) = 0$. However, due to optimality of $J_\infty^*(\cdot)$ we have that $J_\mu(x) \geq J_\infty^*(x)$ for all $x \in \mathcal{X}_\infty$; at the same time it follows from $J_\mu(x) = (\mathbf{T}J_\mu)(x)$ that $J_\mu(x) \leq \ell(x,u) + J_\mu(f_{\mathrm{PWA}}(x,u))$ for all $(x,u) \in \mathcal{X}_\infty \times \mathcal{U}$. Thus from Lemma 7.14 follows $J_\mu(x) \leq J_\infty^*(x)$ for all $x \in \mathcal{X}_\infty$. Therefore, $J_\mu(\cdot) = J_\infty^*(\cdot)$.

(b) For the case $0 \leq J_0(x) \leq J_\infty^*(x)$ for all $x \in \mathcal{X}_\infty$ see [Ber01, Prop. 1.1.5]. In the case $J_0(x) \geq (\mathbf{T}J_0)(x)$ for all $x \in \mathcal{X}_\infty$ we have the following: due to monotonicity of the operator $\mathbf{T}$ [Ber01, Lem. 1.1.1] we have $(\mathbf{T}^k J_0)(x) \geq (\mathbf{T}^{k+1} J_0)(x)$ for all $x \in \mathcal{X}_\infty$ and $k \in \mathbb{N}_{\geq 0}$. Additionally, the sequence $(\mathbf{T}^k J_0)(x(0))$ with increasing k is bounded from below by $J_\infty^*(x(0))$ for all $x(0) \in \mathcal{X}_\infty$, thus the sequence converges to some stationary, bounded solution $J_\mu(\cdot)$, i.e. $J_\mu(x) = (\mathbf{T}J_\mu)(x)$ for all $x \in \mathcal{X}_\infty$. In addition, from part (a) of this theorem we then have $J_\mu(\cdot) = J_\infty^*(\cdot)$.

(c) For an arbitrary feasible input sequence $\{\widetilde{\mu}(x(t))\}_{t=0}^\infty$, due to suboptimality, we clearly have $\infty > J_0(x(0)) \geq J_\infty^*(x(0)) \geq 0$ for all $x(0) \in \mathcal{X}_\infty$. Because $J_0(x(0))$ for all $x(0) \in \mathcal{X}_\infty$ is a realizable cost it follows that

$$J_0(x(0)) = \lim_{T \to \infty} \sum_{t=0}^{T} \ell(x(t), \widetilde{\mu}(x(t))) = \ell(x(0), \widetilde{\mu}(x(0))) + \lim_{T \to \infty} \sum_{t=1}^{T} \ell(x(t), \widetilde{\mu}(x(t)))$$

$$= \ell(x(0), \widetilde{\mu}(x(0))) + J_0(f_{\mathrm{PWA}}(x(0), \widetilde{\mu}(x(0))))$$

$$\geq \min_{u \in \mathcal{U}} \ell(x(0), u) + J_0(f_{\mathrm{PWA}}(x(0), u)) = (\mathbf{T}J_0)(x(0))$$

for all $x(0) \in \mathcal{X}_\infty$. The rest follows directly from part (b) of this theorem.

(a)–(c) From $J_\mu(\cdot) = J_\infty^*(\cdot)$ and the stationarity of $J_\infty^*(\cdot)$ follows directly that $(\mathbf{T}_\mu J_\infty^*) = (\mathbf{T}J_\infty^*)$ and thus with Lemma 7.11 we have that $\mu(x)$ is an optimal control law for all $x \in \mathcal{X}_\infty$. ∎

Corollary 7.16 (Control Lyapunov function). If $J_0(x)$ is continuous at the single point $x = \mathbb{0}_{n_x}$ and chosen according to Theorem 7.15(b) with $J_0(x) \geq (\mathbf{T}J_0)(x)$ for all $x \in \mathcal{X}_\infty$ then it is a (global) control Lyapunov function on $\mathcal{X}_\infty$ for the dynamical system. Moreover, the existence of $J_\mu(\cdot)$ (as defined in (7.16)) is guaranteed, $J_\mu(x) = J_\infty^*(x)$, and $\mu(x) = \mu_\infty^*(x)$ for all $x \in \mathcal{X}_\infty$. □

Proof. Corollary 7.16 follows directly from Theorem 7.15(b) and the definition of a control Lyapunov function, cf. Definition 2.12. ∎

Theorem 7.15 (together with Corollary 7.16) is a rather strong and computationally important result guaranteeing that one will find the unique optimal finitely bounded solution if one exists.

Theorem 7.17 (Convergence rate). Let $J_k(\cdot)$, $\mathcal{X}_k$ be solutions to the DP procedure (7.7a)–(7.7d) such that $\mathcal{X}_k = \mathcal{X}$, $\forall k \geq k^\star$, for some $\mathcal{X} \subseteq \mathbb{R}^{n_x}$ and $k^\star \in \mathbb{N}_{\geq 0}$. Then the *convergence rate*

$$\gamma(k) := \max_{x \in \mathcal{X}} |J_{k+1}(x) - J_k(x)|, \quad k \geq k^\star. \tag{7.18}$$

is a monotonically non-increasing function, i.e. $\gamma(k+1) \leq \gamma(k)$ for all $k \geq k^\star$. □

Proof. Let $k \geq k^* + 1$. From the definition of $J_k(x)$ we see that for all $x \in \mathcal{X}$ the following holds

$$J_k(x) := J_{k-1}(f_{\text{PWA}}(x, \mu_k^*(x))) + \ell(x, \mu_k^*(x)) \tag{7.19a}$$
$$\leq J_{k-1}(f_{\text{PWA}}(x, u)) + \ell(x, u), \qquad \forall (x, u) : f_{\text{PWA}}(x, u) \in \mathcal{X}, \tag{7.19b}$$

where $\mu_k^*(x)$ is the optimal control for a given state $x \in \mathcal{X}$, i.e. it is a function of x. Hence it follows

$$\max_{x \in \mathcal{X}} J_k(x) - J_{k+1}(x)$$
$$= \max_{x \in \mathcal{X}} J_{k-1}(f_{\text{PWA}}(x, \mu_k^*(x))) + \ell(x, \mu_k^*(x)) - J_k(f_{\text{PWA}}(x, \mu_{k+1}^*(x))) - \ell(x, \mu_{k+1}^*(x))$$
$$\leq \max_{x \in \mathcal{X}} J_{k-1}(f_{\text{PWA}}(x, \mu_{k+1}^*(x))) + \ell(x, \mu_{k+1}^*(x))$$
$$\qquad - J_k(f_{\text{PWA}}(x, \mu_{k+1}^*(x))) - \ell(x, \mu_{k+1}^*(x))$$
$$= \max_{x \in \mathcal{X}} \{J_{k-1}(y) - J_k(y) \mid y = f_{\text{PWA}}(x, \mu_{k+1}^*(x))\}$$
$$= \max_{x, y \in \mathcal{X}} \{J_{k-1}(y) - J_k(y) \mid y = f_{\text{PWA}}(x, \mu_{k+1}^*(x))\}$$
$$\leq \max_{y \in \mathcal{X}} J_{k-1}(y) - J_k(y),$$

where the first inequality is due to (7.19b), and the second one due to the removal of some of the constraints from the problem ($x \in \mathcal{X}$ and $y = f_{\text{PWA}}(x, \mu_{k+1}^*(x))$). In a similar way one gets

$$\max_{x \in \mathcal{X}} J_{k+1}(x) - J_k(x)$$
$$= \max_{x \in \mathcal{X}} J_k(f_{\text{PWA}}(x, \mu_{k+1}^*(x))) + \ell(x, \mu_{k+1}^*(x)) - J_{k-1}(f_{\text{PWA}}(x, \mu_k^*(x))) - \ell(x, \mu_k^*(x))$$
$$\leq \max_{x \in \mathcal{X}} J_k(f_{\text{PWA}}(x, \mu_k^*(x))) + \ell(x, \mu_k^*(x)) - J_{k-1}(f_{\text{PWA}}(x, \mu_k^*(x))) - \ell(x, \mu_k^*(x))$$
$$= \max_{x, y \in \mathcal{X}} \{J_k(y) - J_{k-1}(y) \mid y = f_{\text{PWA}}(x, \mu_k^*(x))\}$$
$$\leq \max_{y \in \mathcal{X}} J_k(y) - J_{k-1}(y).$$

Finally, by taking into account that for any function $\tau(x)$ we have

$$\max_x |\tau(x)| = \max \left\{ \max_x \tau(x), \max_x -\tau(x) \right\}.$$

The rest of the proof follows easily with

$$\gamma(k) := \max_{x \in \mathcal{X}} |J_{k+1}(x) - J_k(x)| = \max \left\{ \max_{x \in \mathcal{X}} J_{k+1}(x) - J_k(x), \max_{x \in \mathcal{X}} J_k(x) - J_{k+1}(x) \right\}$$
$$\leq \max \left\{ \max_{x \in \mathcal{X}} J_k(x) - J_{k-1}(x), \max_{x \in \mathcal{X}} J_{k-1}(x) - J_k(x) \right\}$$
$$= \max_{x \in \mathcal{X}} |J_k(x) - J_{k-1}(x)| =: \gamma(k-1). \qquad \blacksquare$$

7.3 CITOC Solution via Dynamic Programming

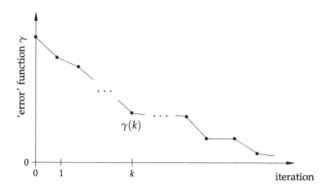

Fig. 7.2: Typical behavior of the 'error' function $\gamma(k)$.

Remark 7.18 (Local convergence). Effectively, Theorem 7.17 states that the DP procedure (7.7a)–(7.7d) does not exhibit *quasi-stationary* behavior (confer Figure 7.2), i.e. it cannot happen that a succession of functions $J_k(\cdot)$ differ only slightly and then suddenly a drastic change appears in $J_{k+1}(\cdot)$ in one additional iteration step of the DP. However, note that functions $J_k(\cdot)$ and $J_{k+1}(\cdot)$ are compared over the whole feasible state space and not for a single point $\tilde{x}$. For a fixed $\tilde{x}$ the *local closureence rate* $\gamma(k, \tilde{x}) = |J_k(\tilde{x}) - J_{k+1}(\tilde{x})|$ is not necessarily a monotonic function. □

As observed in Chapter 6, it may happen that the dynamic program (7.7a)–(7.7d) converges after a finite number of steps, confer Section 6.6.1. Here by convergence is meant that in two successive iterations of the dynamic program the value function (including its domain) does not change, i.e. $J_k(\cdot) = J_{k-1}(\cdot)$ for all $k \geq k_\infty + 1 < \infty$. From this, however, one should in general not assume that the optimal control law steers any feasible state $x(0)$ after at most k_∞ time steps to the origin. Several observations should be made where the 1-/∞-norm CITOC problem may lead to two types of solutions: (a) an optimal control sequence that in a finite number of time steps steers the state to the origin, and (b) an optimal control sequence that takes an infinite number of time steps to steer the state to the origin. This type of behavior may be observed even for constrained linear systems as shown in the following example.

Example 7.19 (Constrained LTI system). Consider the one-dimensional constrained linear time-invariant system

$$\begin{cases} x(t+1) = \tfrac{3}{10}x(t) + u(t) & \text{if } x(t) - 5u(t) \leq 0, \\ x(t) \in [-5, 5], & \text{and} \\ u(t) \in [-1, 1]. \end{cases}$$

The CITOC Problem 7.1 is solved with $Q = 1$, $R = 1$, and $\mathcal{X}_0 = [-5, 5]$ for $p = \infty$.

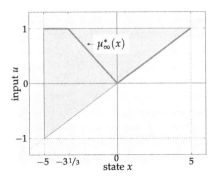
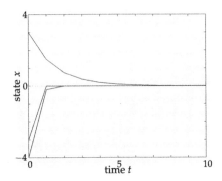

Fig. 7.3: Feasible state-input space and optimal infinite time control law for Example 7.19.

Fig. 7.4: Closed-loop simulation for Example 7.19 for 3 different initial values.

The feasible state space and the optimal infinite time control law are depicted in the extended state-input space in Figure 7.3. If one considers $x(t) \in (0, 5]$ it is obvious that the infinite time optimal control law is in fact $\mu_\infty^*(x(t)) = \frac{1}{5}x(t)$ and therefore the optimal value function or 'cost-to-go' can be computed with $x(t) = (\frac{1}{2})^t x(0)$ as

$$J_\infty^*(x(0)) = \min_{U_\infty} \lim_{T \to \infty} \sum_{t=0}^{T} |x(t)| + |u(t)| = \lim_{T \to \infty} \sum_{t=0}^{T} \frac{6}{5} \cdot (\frac{1}{2})^t |x(0)| = \frac{12}{5}x(0).$$

From this it can be seen that for any state starting in $(0, 5]$ it will take an infinite number of steps until the state reaches the origin; whereas if the state $x(t) \in \left[\frac{-10}{3}, 0\right]$ the infinite time optimal control law is given by $\mu_\infty^*(x(t)) = -\frac{3}{10}x(t)$ and thus drives the state in one single step to the origin.

With similar considerations for the rest of the feasible state space one finds for the infinite time optimal control law

$$\mu_\infty^*(x(t)) = \begin{cases} \frac{1}{5}x(t) & \text{if } x(t) \in (0, 5], \\ -\frac{3}{10}x(t) & \text{if } x(t) \in \left[\frac{-10}{3}, 0\right], \\ 1 & \text{if } x(t) \in \left[-5, \frac{-10}{3}\right) \end{cases}$$

and for the cost-to-go

$$J_\infty^*(x(0)) = \begin{cases} \frac{12}{5}x(0) & \text{if } x(0) \in (0, 5], \\ -\frac{13}{10}x(0) & \text{if } x(0) \in \left[\frac{-10}{3}, 0\right], \\ -\frac{139}{100}x(0) - \frac{3}{10} & \text{if } x(0) \in \left[-5, \frac{-10}{3}\right). \end{cases}$$

The optimal infinite time closed-loop trajectory for three different initial values is depicted in Figure 7.4.

7.3.2 Stabilizing Suboptimal Control

In contrast to the 'traditional' initialization method of the dynamic program, where $J_0(\cdot) \equiv 0$, one can obtain stabilizing suboptimal controllers even in the intermediate iteration steps of the DP as is stated in the following theorem.

Theorem 7.20 (Stabilizing suboptimal control). Let Assumption 7.2 hold and let $J_k : \mathcal{X}_\infty \to \mathbb{R}_{\geq 0}$ be some arbitrary finitely bounded function with $J_k(\mathbb{0}_{n_x}) = 0$ and $J_k(\cdot)$ continuous at $x = \mathbb{0}_{n_x}$. If $J_k(x) \geq (\mathbf{T}J_k)(x)$ for all $x \in \mathcal{X}_\infty$ then

$$\widetilde{\mu}(x) := \arg\min_{u \in \mathcal{U}} \ell(x,u) + J_k(f_{\text{PWA}}(x,u))$$

is a globally asymptotic stabilizing (suboptimal) control law for all $x \in \mathcal{X}_\infty$. Moreover, $J_k(\cdot)$ is a global Lyapunov function for the controlled system when the control law $\widetilde{\mu}(\cdot)$ is applied. □

Proof. First, from $J_k(x) \geq (\mathbf{T}J_k)(x) = \ell(x,\widetilde{\mu}(x)) + J_k(f_{\text{PWA}}(x,\widetilde{\mu}(x)))$ follows directly that $-\Delta J_k(x) := J_k(x) - J_k(f_{\text{PWA}}(x,\widetilde{\mu}(x))) \geq \ell(x,\widetilde{\mu}(x)) \geq \|Qx\|_p$ for all $x \in \mathcal{X}_\infty$. Because Q is of full column rank, there always exists a finite $\alpha_1 > 0$ with $\|Qx\|_p \geq \alpha_1 \|x\|_2$ and thus $-\Delta J_k(x) \geq \alpha_1 \|x\|_2$ for all $x \in \mathcal{X}_\infty$. This means that $-\Delta J_k(\cdot)$ is always bounded below by some K-class function. Second, with $J_k(\cdot) \geq 0$ it follows $J_k(x) \geq (\mathbf{T}J_k)(x) \geq \min_{u \in \mathcal{U}} \ell(x,u) \geq \|Qx\|_p \geq \alpha_2 \|x\|_2$ for some finite $\alpha_2 > 0$. By similar argument as in Lemma 7.4(c) there exists a K-class function $\bar{J}(\cdot)$ bounding $J_k(\cdot)$ from above. From these three statements it follows that $J_k(\cdot)$ is a global Lyapunov function [Vid93, Laz06] on $\mathcal{X}_\infty$ for the closed-loop system and thus the stability argument of $\widetilde{\mu}(\cdot)$ follows directly. ∎

This result is somewhat intuitive and in the line of Corollary 7.16 but at the same time very interesting: it allows one to compute in an 'easy' way a stabilizing controller $\widetilde{\mu}(\cdot)$ as well as a bound $(\mathbf{T}J_k)(\cdot)$ on the optimality/performance of the controller. Simultaneously a Lyapunov function, $J_k(\cdot)$, for the controlled system is given. Moreover, this result shows that if the value function iteration is started with some $J_0(\cdot)$ (confer e.g. Theorem 7.15(a)) and at some iteration k of the dynamic program it is detected that $J_k(\cdot) \geq (\mathbf{T}J_k)(\cdot)$, then in all the following iteration steps of the dynamic program stabilizing controllers and continously improving optimality bounds are computed. For initial value functions $J_0(\cdot)$, which fulfill the conditions of the second case of Theorem 7.15(b) or Theorem 7.15(c), this is already true from the beginning, i.e. for all $k \geq 0$.

Please note that for initial value functions $J_0(\cdot)$ which fulfill the classical convergence conditions, i.e. $0 \leq J_0(\cdot) \leq J^*_\infty(\cdot)$, of Theorem 7.15(b) this stabilization and optimality bound property can only be given after computing the limit function $J^*_\infty(\cdot)$ but not for the intermediate iteration steps. In fact, during the classical iteration from 'below', destabilizing controllers are likely to be computed, as shown in the following simple Example 7.21.

Example 7.21 (DP iteration starting from 'below' and 'above'). Consider the simple one-dimensional unconstrained time-invariant system

$$x(t+1) = 2x(t) + u(t), \quad \text{where} \quad x(t) \in \mathbb{R} \text{ and } u(t) \in \mathbb{R}.$$

It is easy to see that

$$u(t) = \mu_{\text{stab}}(x(t)) = \alpha x(t) \quad \text{with} \quad \alpha \in (-3, -1)$$

is a set of stabilizing controllers for the system and $u(t) = \mu(x(t)) = \alpha x(t)$ with $\alpha \in \mathbb{R} \setminus [-3, -1]$ are destabilizing. If one wants to solve the CITOC Problem 7.1 for $Q = 1$, $R = 1$, $\mathcal{X}_0 = \mathbb{R}$, and $p \in \{1, \infty\}$, and utilize the classical dynamic programming iteration from 'below', the following steps are computed:

$J_0(x(0)) \equiv 0,$
$J_1(x(0)) = |x(0)|$ with $u(t) = \mu_1(x(t)) \equiv 0,$
$J_2(x(0)) = 3|x(0)|$ with $u(t) = \mu_2(x(t)) = \tilde{\alpha}x(t),\ \tilde{\alpha} \in [-2, 0],$
$J_3(x(0)) = J^*_\infty(x(0)) = 3|x(0)|$ with $u(t) = \mu_3(x(t)) = \mu^*_\infty(x(t)) = -2x(t).$

Note that $\mu_1(\cdot)$ and $\mu_2(\cdot)$ with $\tilde{\alpha} \in (-1, 0]$ are destabilizing controllers for the system. Clearly, if one initializes from below with e.g. $J_0(x(0)) = |x(0)|$ one also computes destabilizing controllers for the system. This is in contrast to if the dynamic program is started from 'above' where in every DP iteration stabilizing controllers are computed:

$J_0(x(0)) = \beta|x(0)|,$ where $\beta \geq 3,$
$J_1(x(0)) = 3|x(0)|,$ with $u(t) - \mu_1(x(t)) = 2x(t),$
$J_2(x(0)) = J^*_\infty(x(0)) = 3|x(0)|,$ with $u(t) = \mu_2(x(t)) = \mu^*_\infty(x(t)) = -2x(t).$

7.4 An Efficient Algorithm for the CITOC Solution

In this section, first it is described how the properties of the solution to the rather general DP problem (7.7a)–(7.7d) can be exploited and used for an efficient computation of the infinite time solution. Subsequently, the specific implementation of the algorithm is presented.

In the rest of this chapter for some set or function τ, the set or function $\mathring{\tau}$ denotes its restriction to the neighborhood of the origin $[x'\ u']' = \mathbb{0}_{n_x+n_u}$. For instance, $\mathring{\mathcal{D}}$ describes the domain of those PWA dynamics that are valid for the origin

$$\{\mathring{\mathcal{D}}_d\}_{d=1}^{N_{\mathring{\mathcal{D}}}} := \{\mathcal{D}_i \in \mathcal{D} \mid \mathbb{0}_{n_x+n_u} \in \mathcal{D}_i \text{ with } i = 1, \ldots, N_{\mathcal{D}}\},$$

while $\mathring{f}_{\text{PWA}}(\cdot, \cdot)$ represents the restriction of the function $f_{\text{PWA}}(\cdot, \cdot)$ to that domain. Furthermore, equivalently to Section 6.3, when it is mentioned to SOLVE

7.4 An Efficient Algorithm for the CITOC Solution

iteration k of a DP, it is meant to formulate several multi-parametric linear programs for it and obtain a triplet of expressions for the value function, the control law (optimizer), and the polyhedral partition of the feasible state space

$$\left(J_k(\cdot),\ \mu_k(\cdot),\ \{\mathcal{P}_{k,i}\}_{i=1}^{N_k}\right), \tag{7.20}$$

cf. Algorithm 7.22 and 7.23. By inspection of the DP problem (7.7a)–(7.7d) one sees that at each iteration step $N_\mathcal{D}\, N_{k-1}$ mp-LPs are solved. After that, by using polyhedral manipulation one has to compare all generated regions, check if they intersect and remove the redundant ones, before storing a new partition that has N_k regions.

Under the Assumptions 4.2 and 7.2, all closed-loop trajectories that converge to the origin $\mathbb{0}_{n_x+n_u}$ in an infinite number of time steps have to go through some of the PWA dynamics associated with the domain $\mathring{\mathcal{D}}_d$, $d = 1, \ldots, \mathring{N}_\mathcal{D}$, and regions $\mathring{\mathcal{P}}_{\infty,i}$, $i = 1, \ldots, \mathring{N}_\infty$, that are touching the origin, cf. Example 7.19. Thus at the beginning, instead of focusing on the whole feasible state space and the whole domain of the system, one can limit the algorithm to the neighborhood of the origin and only after attaining convergence one proceeds with the exploration of the rest of the state space. In this way at each iteration step of the DP one would – on average – have to solve a much smaller number of mp-LPs. Let us call the solution to such a restricted problem the 'core' $\mathcal{C}_0$. Note that in general the core $\mathcal{C}_0$ is a non-convex polyhedral collection.

Any positively invariant set is a valid candidate for the core $\mathcal{C}_0$, as long as an associated control strategy is feasible and steers the state to the origin. The only prerequisite is that for any given initial feasible state, i.e. the state for which the original problem has a solution, one can reach at least one element of the core in a finite number of time steps. However, as its name says, the core is used as a 'seed' or starting point for the future construction and exploration of the feasible state space. Thus, obtaining a good suboptimal solution for the core is desirable in order to limit the number of iterations which improve this very part of the state space in further DP iterations.

The task of solving the CITOC problem (7.3a)–(7.3b) is split into two subproblems and respective algorithms. In the first algorithm (Algorithm 7.22) the portion of the state space around the origin is explored and the 'core' of the infinite time solution is constructed. In the second algorithm (Algorithm 7.23), starting from the core $\mathcal{C}_0$, a sequence of additions – named '*rings*' $\mathcal{R}_k$ – to the core $\mathcal{C}_{k-1}$ are build until the algorithm converges for the whole feasible state space $\mathcal{X}_\infty$. At the end one obtains the infinite time solution $\mathcal{S}_\infty^*$. Here with $\mathcal{C}_k$, $\mathcal{R}_k$, and $\mathcal{S}_k$ the triplets of the form given in (7.20) is denoted.

In an ideal scenario the core $\mathcal{C}_0$ would be part of an infinite time optimal solution, and every ring $\mathcal{R}_k$ would also be a part of an infinite time solution. Then in all intermediate steps we would have to explore only the one step ahead optimal transitions from the whole domain of the PWA dynamics to the latest ring (instead of going from the whole domain of the dynamics to the initial core and all previous rings). In practice one is likely to observe sub-ideal scenarios: the newly generated ring, $\mathcal{R}_k$, may contain polyhedra with associated value functions that are 'worse' (meaning bigger) than the infinite time solution and thus such polyhedra will be altered in the future steps of the algorithm.

As stated in Section 6.3, the function INTERSECT & COMPARE removes such polyhedra that are completely covered with other polyhedra [Bao05, KGB04] which have a corresponding 'better' (meaning smaller) value function expression. If some polyhedron $\mathcal{P}_{k,i}$ is only partially covered with better 'regions' the part of $\mathcal{P}_{k,i}$ with the smaller cost can be partitioned into a set of convex polyhedra. Thus the polyhedral nature of the feasible state space partition is preserved in each iteration of Algorithm 7.22. Note that in the SOLVE step one is solving a smaller number of problems than in the general DP. Since we are restricting ourselves to the neighborhood of the origin, the number of regions at each step is likely to remain rather small. However, the choice of the initial $\mathring{J}_0(\cdot) \equiv 0$ may lead to a big number of iterations depending on the desired precision. If a better or other initial guess for $\mathring{J}_0(\cdot)$ is known, confer Section 7.4.1, it can be used to speed up Algorithm 7.22.

Algorithm 7.22 (Generating the CORE of the infinite time solution) _____

INPUT $\mathring{f}_{\text{PWA}}(x,u)$, $\{\mathring{\mathcal{D}}_d\}_{d=1}^{\mathring{N}_\mathcal{D}}$, p, Q, R, $k_{\max}$
OUTPUT The core $\mathcal{C}_0$

LET $\mathring{\mathcal{S}}_0 \leftarrow (\mathring{J}_0(x) := 0,\ \mathring{\mu}_0(x) := \mathbb{0}_{n_u},\ \mathring{\mathcal{P}}_{0,1} := \mathcal{X}_0)$
LET $k \leftarrow 0$, finished $\leftarrow$ **false**
WHILE $k < k_{\max}$ **AND NOT** finished
   LET $k \leftarrow k+1$
   FOR $d = 1$ **TO** $\mathring{N}_\mathcal{D}$
     FOR EACH $\mathring{\mathcal{P}}_{k-1,i} \in \mathring{\mathcal{S}}_{k-1}$

$$s_{d,i} \leftarrow \text{SOLVE}\ \min_u \|Qx\|_p + \|Ru\|_p + \mathring{J}_{k-1}(\mathring{f}_{\text{PWA}}(x,u))$$
$$\text{subj. to}\ \begin{cases} [x',u']' \in \mathring{\mathcal{D}}_d, \\ \mathring{f}_{\text{PWA}}(x,u) \in \mathring{\mathcal{P}}_{k-1,i} \end{cases}$$

     END
   END
   LET $\mathcal{S}_k \leftarrow$ INTERSECT & COMPARE $\{s_{d,i}\}$
   LET $\mathring{\mathcal{S}}_k \leftarrow$ RESTRICTION of $\mathcal{S}_k$ to the origin
   IF $\mathring{\mathcal{S}}_k = \mathring{\mathcal{S}}_{k-1}$
     THEN finished $\leftarrow$ **true**, $\mathcal{C}_0 \leftarrow \mathring{\mathcal{S}}_k$
END

7.4 An Efficient Algorithm for the CITOC Solution

Note that C_0 is a positively invariant set by construction. After constructing the initial core, C_0, one can proceed with the exploration of the rest of the state space as described in the following Algorithm 7.23.

Algorithm 7.23 (Generating the infinite time solution)

INPUT $f_{PWA}(x,u)$, $\{D_d\}_{d=1}^{N_D}$, C_0, p, Q, R, k_{max}
OUTPUT The infinite time solution S_∞^*

LET Solution $S_0 \leftarrow C_0$
LET Ring $\mathcal{R}_0 \leftarrow C_0$
LET $k \leftarrow 0$, finished $\leftarrow$ **false**
WHILE $k < k_{max}$ **AND NOT** finished
  LET $k \leftarrow k+1$
  FOR $d = 1$ **TO** N_D
    FOR EACH $\mathcal{P}_{k-1,i} \in \mathcal{R}_{k-1}$

$$s_{d,i} \leftarrow \text{SOLVE} \min_u \|Qx\|_p + \|Ru\|_p + J_{k-1}(f_{PWA}(x,u)),$$
$$\text{subj. to } \begin{cases} [x',u']' \in \mathcal{D}_d, \\ f_{PWA}(x,u) \in \mathcal{P}_{k-1,i} \end{cases}$$

    END
  END
  LET $S_k \leftarrow$ INTERSECT & COMPARE $S_{k-1}, \{s_{d,i}\}$
  LET $C_k \leftarrow S_k \cap S_{k-1}$
  LET $\mathcal{R}_k \leftarrow S_k \backslash C_k$
  IF $\mathcal{R}_k = \emptyset$
    THEN finished $\leftarrow$ **true**, $J_\infty^*(x) \leftarrow J_k(x)$, $S_\infty^* \leftarrow S_k$
END

Note that if Algorithm 7.22 ends successfully and if any optimal closed-loop trajectory starting in C_0 happens to stay in C_0 for all time then the value function computed in Algorithm 7.22 associated with C_0 is in fact the optimal value function $J_\infty^*(\cdot)$ associated with the feasible set of C_0. However, if any optimal closed-loop trajectory starting in C_0 leaves this set in finite time then the value function computed in Algorithm 7.22 associated with C_0 is the best current upper bound of the optimal value function $J_\infty^*(x(0))$ for all $x(0) \in C_0$. In the case of the later scenario, by optimality, Algorithm 7.23 will account for this fact and improve the value function in the corresponding part of C_0 by going through additional DP iterations until optimality is reached. The same argument holds for all the computed intermediate 'cores', C_k, of Algorithm 7.23 as well.

7.4.1 Alternative Choices of the Initial Value Function

Theorem 7.15 guarantees the convergence of the dynamic programming based algorithm to the optimal solution $J_\infty^*(\cdot)$ starting from an (almost) arbitrary initial value function $J_0(\cdot)$. This can be used in order to decrease or limit the number of iterations needed in the proposed algorithm. Moreover, at the same time an upper bound to the optimal solution can be given.

From the above discussion we know that $J_\infty^*(0_{n_x}) = 0$ as well as $\mu_\infty^*(0_{n_x}) = 0_{n_u}$; and for some set $\mathring{\mathcal{X}} \subseteq \mathcal{X}_\infty$ around the origin the optimal value function $J_\infty^*(\cdot)$ and the optimal control law $\mu_\infty^*(\cdot)$ is a piecewise *linear* function of the state x.

In Algorithm 7.22 the exploration and computation is limited to such regions around the origin and only the domain of the system dynamics that touch the origin or have the origin in the interior are considered.

Now, consider the special case of system (4.1) when $N_D = 1$, i.e. the linear, stabilizable system

$$x(t+1) = Ax(t) + Bu(t) \tag{7.21}$$

with polyhedral state and input constraints $\mathcal{D} = \mathcal{X} \times \mathcal{U}$. In addition, consider for this system the stabilizing state feedback control law for states $x \in \mathring{\mathcal{X}}$ around the origin

$$\mu(x) = Kx \quad \text{with} \quad \|A^{\text{CL}}\|_p < 1, \tag{7.22}$$

where $A^{\text{CL}} := A + BK$ and $\|A^{\text{CL}}\|_p$ with $p \in \{1, \infty\}$ denotes the standard Hölder matrix 1- or ∞-norm [HJ85] of A^{CL}, cf. Definition 1.40.

J_0 as an Upper Bound to J_∞^* by Approximation

Then for control law (7.22) one has for all $x(0) \in \mathring{\mathcal{X}}$ that

$$J_\infty^*(x(0)) \leq J_\infty(x(0), u(t) = Kx(t))$$
$$= \lim_{T \to \infty} \sum_{t=0}^{T} \|Q(A^{\text{CL}})^t x(0)\|_p + \|RK(A^{\text{CL}})^t x(0)\|_p$$
$$\leq \lim_{T \to \infty} \sum_{t=0}^{T} (\|Q\|_p + \|RK\|_p) \cdot \|(A^{\text{CL}})^t\|_p \cdot \|x(0)\|_p$$
$$\leq (\|Q\|_p + \|RK\|_p) \|x(0)\|_p \cdot \lim_{T \to \infty} \sum_{t=0}^{T} \|A^{\text{CL}}\|_p^t$$
$$= \left\| \frac{\|Q\|_p + \|RK\|_p}{1 - \|A + BK\|_p} \cdot I_{n_x} \cdot x(0) \right\|_p =: J_0(x(0)).$$

Clearly, for any bounded $x(0)$ it follows that $\infty > J_0(x(0)) \geq J_\infty^*(x(0))$.

7.4 An Efficient Algorithm for the CITOC Solution

Note that this choice of $J_0(\cdot)$ is of a very simple type and might serve as a 'good' first try for the DP iteration procedure but does not necessarily belong to the class of functions which are covered by Theorem 7.15.

However, finding a tight bound $J_0(x(0))$ on $J_\infty^*(x(0))$ for all $x(0) \in \mathring{\mathcal{X}}$, i.e

$$\min_K \quad \frac{\|Q\|_p + \|RK\|_p}{1 - \|A + BK\|_p} \quad (7.23a)$$

$$\text{subj. to} \quad \|A + BK\|_p < 1, \quad (7.23b)$$

might be a difficult nonlinear or even infeasible problem due to condition (7.23b), depending on the system data.

J_0 as an Upper Bound to J_∞^* by Lyapunov Function Construction

For a given stabilizing state feedback control law $\mu(x) = Kx$ for system (7.21), we can always compute a Lyapunov function for the closed-loop system of the type $V(x) = \|Wx\|_p$ with $p \in \{1, \infty\}$, $W \in \mathbb{R}^{m \times n_x}$, $\text{rank}(W) = n_x$, $\infty > m \geq n_x$ as proposed in Chapter 8 and [Pol95, Pol97, CM07]. In order to guarantee that such a Lyapunov function is always an upper bound to $J_\infty^*(\cdot)$, i.e. $J_\infty^*(\cdot) \leq J_0(\cdot)$, the scaling

$$J_0(x) := \alpha_W \|Wx\|_p \quad (7.24a)$$

with $\alpha_W > 0$ needs to be performed. From Chapter 8 follows that a pair $(W, Z) \in \mathbb{R}^{m \times n_x} \times \mathbb{R}^{m \times m}$ exists which fulfills $WA^{CL} = ZW$ and $\|Z\|_p < 1$; and from Section 8.9, cf. also Algorithm 8.12, it follows

$$\alpha_W \geq \frac{\|\tilde{P}(W'W)^{-1}W'\|_p}{1 - \|Z\|_p}, \quad (7.24b)$$

where $\tilde{P} := \begin{bmatrix} Q \\ RK \end{bmatrix}$ for $p = 1$. For the case $p = \infty$ one can chose $\tilde{P} := 2\begin{bmatrix} Q \\ RK \end{bmatrix}$. (A tight bound $\|\tilde{P}x\|_\infty := \|Qx\|_\infty + \|RKx\|_\infty$ for $p = \infty$ can be found via one simple mp-LP.)

$J_0(\cdot)$ of the type described by Equation (7.24) fulfills the prerequisites of Corollary 7.16 and thus using it as initial value function of the dynamic programming iteration guarantees convergence to the optimal solution.

J_0 as an Upper Bound to J_∞^* by Lyapunov Function Construction for Piecewise Affine Systems

A similar approach as described above, where a Lyapunov function of the type $\|Wx\|_p$ is scaled in a particular manner, can also be applied for a class of piecewise affine systems. See Section 8.9 for further detail.

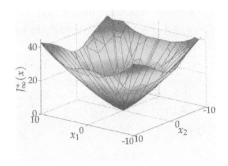
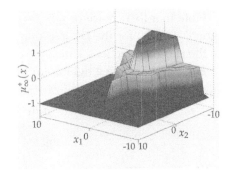

Fig. 7.5: Value function $J_\infty^*(\cdot)$ of the infinite time solution S_∞^* for system (6.16). Same coloring implies the same cost value.

Fig. 7.6: PWA optimal control law $\mu_\infty^*(\cdot)$ of the infinite time solution S_∞^* for system (6.16).

Another, rather involved, Lyapunov function construction for the considered class of PWA systems (4.1) is described in [LHW+05]. However, one would still need to scale the computed Lyapunov function $V(\cdot)$ in a similar way as described in previously in order to guarantee that the obtained $J_0(\cdot)$ is an upper bound to the optimal solution $J_\infty^*(\cdot)$. Again, this $J_0(\cdot)$ fulfills the prerequisites of Corollary 7.16 and thus serves as initial value function of the dynamic programming iteration and guarantees convergence to the optimal solution.

7.5 Examples

7.5.1 Constrained PWA Sine-Cosine System

Consider once more the constrained piecewise affine system [BM99a] from Section 6.6.1:

$$\begin{cases} x(t+1) = \frac{4}{5}\begin{bmatrix} \cos\alpha(x(t)) & -\sin\alpha(x(t)) \\ \sin\alpha(x(t)) & \cos\alpha(x(t)) \end{bmatrix} x(t) + \begin{bmatrix} 0 \\ 1 \end{bmatrix} u(t), \\ \alpha(x(t)) = \begin{cases} \frac{\pi}{3} & \text{if } [1\ 0]x(t) \geq 0, \\ -\frac{\pi}{3} & \text{if } [1\ 0]x(t) < 0, \end{cases} \\ x(t) \in [-10,10] \times [-10,10], \text{ and } u(t) \in [-1,1]. \end{cases} \quad (6.16)$$

The CITOC Problem 7.1 is solved with $Q = I_2$, $R = 1$, and $\mathcal{X}_0 = [-10,\ 10] \times [-10,\ 10]$ for $p = \infty$.

As described in Section 7.4 the algorithm is divided into two parts: first the so called inner core $\mathcal{C}_0$ is constructed via a dynamic programming approach (Algorithm 7.22). After this inner core has converged it serves as a 'seed' (or optimal current upper bound of the value function restricted to the part of the state space in this particular step) for the second part of the algorithm

7.5 Examples

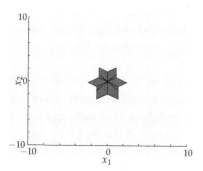

(a) Value function of the 'inner core' $\mathcal{C}_0$.

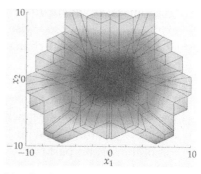

(b) Value function at the intermediate construction step $k = 4$.

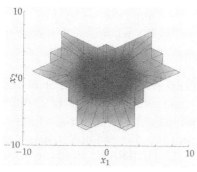

(c) Core of the value function at the intermediate construction step $k = 4$.

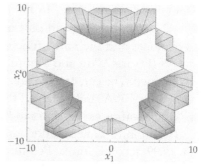

(d) Ring of the value function at the intermediate construction step $k = 4$.

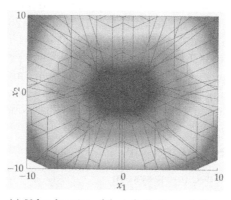

(e) Value function of the infinite time solution.

Fig. 7.7: State space partition of the value function and of the piecewise affine control law of system (6.16). Same coloring in Figure (a)–(e) implies the same cost value.

(Algorithm 7.23), where from the seed the rest of the feasible state space is explored until the piecewise affine value function for the whole feasible state space $\mathcal{X}_\infty$ does not change for two successive steps in the exploration procedure.

For system (6.16) the inner core, $\mathcal{C}_0$, is computed in 5 iteration steps. Figure 7.7(a) shows the state space partitioning comprising 10 polyhedral regions of the value function of the inner core. Figure 7.7(b) shows the state space partition of the value function with 188 polyhedral regions at the intermediate step $k = 4$ of the second part of the algorithm. Figure 7.7(c) shows the state space partition $\mathcal{C}_4$ of the current optimal upper bound of the infinite time value function which does not change from the intermediate step $k = 3$ to $k = 4$ and it consists of 104 regions. This can be viewed as the new core in step $k = 4$ from which the ring $\mathcal{R}_4$ in Figure 7.7(d) was computed in step $k = 4$. After $k = 7$ steps of the second part of the algorithm the whole feasible state space is explored and the value function $J_k(\cdot)$ does not change from step $k = 7$ to $k = 8$ (Figure 7.7(e) and Figure 7.5) and thus the infinite time solution $J_\infty^*(\cdot)$ is obtained. The constructed state space partition consists of 252 polyhedral regions. The PWA control law of the infinite time solution consists of 23 different piecewise affine expressions and is depicted in Figure 7.6.

The infinite time solution to (7.3a)–(7.3b) for this example was obtained in 49.99 seconds on a Pentium M, 1.6 GHz (1GB RAM) machine running MATLAB® 7.0.4 SP2, the LP solver of NAG® [Num02], and MPT 2.6.1 [KGB04, KGBM03]. This shows the efficiency of the proposed algorithm compared to the approach given in Chapter 6 or [BCM03], where the 'brute force' computation of the infinite time solution with $T = 11$ took 288.78 seconds on the same machine. Moreover, it should be mentioned that in order to obtain the infinite time solution using the CFTOC algorithm one needs to perform the rather computationally expensive operation to test and compare always two consecutive value functions during the iteration procedure, to know when to terminate the algorithm, which is not accounted for in the 288.78 seconds.

7.5.2 Constrained LTI System

Consider the constrained linear system

$$\begin{cases} x(t+1) = \begin{bmatrix} 7/10 & -1/10 & 0 \\ 1/5 & -1/2 & 1/10 \\ 0 & 1/10 & 1/10 \end{bmatrix} x(t) + \begin{bmatrix} 1/10 & 0 \\ 1/10 & 1 \\ 1/10 & 0 \end{bmatrix} u(t), \\ -5\mathbb{1}_3 \leq x(t) \leq 5\mathbb{1}_3, \quad \text{and} \quad -5\mathbb{1}_2 \leq u(t) \leq 5\mathbb{1}_2 \end{cases} \quad (7.25)$$

for which the CITOC Problem 7.1 is solved with $P = 0_{3\times 3}$, $Q = I_3$, and $R = \frac{1}{10}I_2$ for $p = \infty$.

7.5 Examples

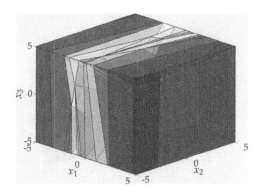

Fig. 7.8: Polyhedral partition $\{\mathcal{P}_i\}_{i=1}^{80}$ of the infinite time solution for example (7.25).

For system (7.25) the inner core, $\mathcal{C}_0$, comprising 44 polyhedral regions is computed in 16 iteration steps. The infinite time solution to (7.3a)–(7.3b) for this example was obtained in 139.46 seconds on a Pentium M, 1.6 GHz (1GB RAM) machine running MATLAB® 7.0.4 SP2, the LP solver of NAG® [Num02], and MPT 2.6.1 [KGB04, KGBM03] and is defined over a polyhedral partition with only 80 regions, cf. Figure 7.8.

The infinite time solution using the CFTOC dynamic programming approach given in Chapter 6 can be obtained for a horizon of $T = T_\infty = 13$ and took 942.15 seconds on the same machine setup as above. Again, this computation time corresponds to a pure computation of the CFTOC solution with that horizon without intermediate comparisons of the value functions in order to detect if the infinite horizon solution is obtained. For a fair timing comparison these intermediate comparison computations should be added to the 942.15 seconds.

7.5.3 Car on a PWA Hill

Recall the example from Section 6.6.2 (page 61ff) in which a car was moving horizontally on a piecewise affine 'environment'. The goal of the car is to climb to the top of a steep hill and then to maintain its position at the top (the origin), without falling from the piecewise affine environment. The discrete-time model is given by the following constrained and discontinuous piecewise affine system

$$x(t+1) = \begin{bmatrix} 1 & 1/2 \\ 0 & 1 \end{bmatrix} x(t) + \begin{bmatrix} 1/8 \\ 1/2 \end{bmatrix} u(t) + a(x(t)), \quad (6.18a)$$

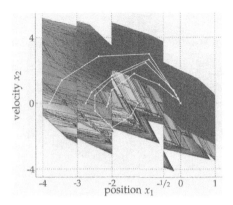

Fig. 7.9: State-space partition of the infinite time optimal closed-form solution $\mu_\infty^*(\cdot)$ for the example in Section 7.5.3. The coloring corresponds to value function $J_\infty^*(\cdot)$. Trajectories indicate the stable closed-loop behavior.

where

$$a(x(t)) = \begin{cases} 0_2, & \text{if } [1\ 0]x(t) \in [-\tfrac{1}{2},\ 1], \\ -\tfrac{1}{4}g\sin(20\tfrac{\pi}{180})\begin{bmatrix}1\\2\end{bmatrix}, & \text{if } [1\ 0]x(t) \in [-2,\ -\tfrac{1}{2}], \\ 0_2, & \text{if } [1\ 0]x(t) \in [-3,\ -2], \\ -\tfrac{1}{4}g\sin(-5\tfrac{\pi}{180})\begin{bmatrix}1\\2\end{bmatrix}, & \text{if } [1\ 0]x(t) \in [-4,\ -3], \end{cases}$$

g is the gravitational constant, the first coordinate of x, i.e. x_1, is the horizontal position of the car, and the second, x_2, is its horizontal velocity. Moreover, the control action is constrained by

$$|u(t)| \leq 2 \quad \text{and} \quad |u(t+1) - u(t)| \leq 40$$

which prohibits the car to directly climbing up the steep hill in $x_1 \in [-2, -\tfrac{1}{2}]$.

The CITOC Problem 7.1 is solved with $Q = \text{diag}([100, 1]')$, $R = 5$, and $\mathcal{X}_0 = [-4, 1] \times [-40, 40]$ for $p = 1$ in 6.9 hours (24 864 seconds) on a Pentium 4, 2.8 GHz machine on Linux running MATLAB® 7.1.0 SP3, the LP solver of NAG® [Num02], and MPT 2.6.1 [KGB04, KGBM03].

The state-space partition of the infinite time optimal closed-form control solution $u(t) = \mu_\infty^*(x(t))$ is depicted in Figure 7.9 and comprises 4034 polyhedral regions. The coloring corresponds to the value function $J_\infty^*(\cdot)$ and ranges from dark blue to dark red, where dark blue refers to the value zero which is naturally obtained at the origin, cf. also Figure 7.10. Moreover, the coloring indicates the 'flow' (cost) of the closed-loop trajectories from dark red to dark blue. Sample trajectories show the stable closed-loop behavior.

As mentioned before, observe that simply 'climbing' the hill, starting on the hill or at the bottom of the hill, is not possible. The optimal control strategy for

7.5 Examples

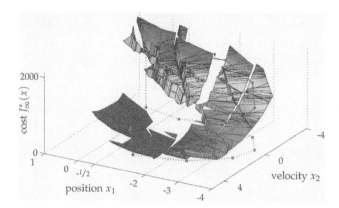

Fig. 7.10: Infinite time optimal value function $J^*_\infty(\cdot)$ for the example in Section 7.5.3. Trajectory indicates the stable closed-loop behavior.

initial states in these positions with zero initial velocity, drives the car first to the left in order to gain enough speed to cope with the steep slope.

Part III

Analysis and Post-processing Techniques for Piecewise Affine Systems

8

LINEAR VECTOR NORMS AS LYAPUNOV FUNCTIONS

This chapter continues on the results proposed in [Pol95, KAS92] for computing vector 1- or ∞-norm based Lyapunov functions for linear discrete-time systems by presenting an algorithm with guaranteed finite termination for the construction of Lyapunov functions of that class. The algorithm utilizes the solution of a finite sequence of feasibility LPs with very few constraints or, equivalently, very simple algebraic tests in contrast to solving the original problem where a bilinear matrix equation with rank and norm constraint needs to be solved. Such Lyapunov functions are utilized for a priori imposing closed-loop stability and feasibility for all time when using a receding horizon control policy. A simple algorithm for this purpose for a class of piecewise affine systems is presented.

8.1 Introduction

Lyapunov functions are the most successful tool, both in stability analysis and in the heart of many 'advanced' control synthesis techniques for linear as well as for nonlinear systems. For instance, one of many important applications of Lyapunov functions is in the framework of finite horizon optimal control (cf. Chapter 6) such as *receding horizon control* (cf. Chapter 3), where Lyapunov functions are used in the cost function to upper bound the infinite horizon cost in order to a priori guarantee asymptotic stability and feasibility for all time (Definition 3.4) of the closed-loop system.

In accordance to energy consumption arguments in a system, the class of quadratic functions are most often used as Lyapunov function candidates. However, as argued in Section 3.3, in many (real life) situations when the aforementioned cost function of the optimal control problem is composed of weighted linear vector norms the class of quadratic Lyapunov functions is unsuitable, cf. for example [KA02, BCM06, LHW+05, SX97, Bor03, Erd04, LL06, TW01, EM99, RH05, BBM02].

Here, we consider real valued, finitely parameterized vector 1- or ∞-norm based Lyapunov functions for discrete-time systems. The idea of using vector norm Lyapunov functions by scaling faces of a polytope was introduced in [MS72] and it was shown in [KAS92, Pol95] that for linear systems a 1- or ∞-norm based Lyapunov function exists if and only if the system is stable. The origin of the modified existence proof goes back to [MP86] which, however, is non-constructive. A constructive proof using linear programming (LP) for linear continuous-time systems is presented in [Pol97] for ∞-norm based Lyapunov functions. A direct optimization based search for a Lyapunov function of the considered type is in general very difficult and limited to low dimensional systems, cf. for example Section 8.10. The authors in [KAS92, Bit88] presented an algebraic construction of an ∞-norm based Lyapunov function for the case when the system has distinct eigenvalues located in the rhombus spanned by the vertices $1, +j, -1, -j$ in the complex plane. [Pol95] gives a more general insight for the Jordan decomposition based construction of real valued ∞-norm based Lyapunov functions for continuous-time systems. However, the detailed construction for continuous-time systems having multiple eigenvalues or for discrete-time systems is left to the reader.

Even though similarities exist, the formulation does not straightforwardly translate to the discrete-time case as it will be shown in the following.

Here a general algorithm with guaranteed finite termination for the construction of weighted vector 1- and ∞-norm based Lyapunov functions is presented. The algorithm utilizes the solution of a finite sequence of feasibility LPs with few constraints (Section 8.5) or, equivalently, simple algebraic tests (Section 8.6) in contrast to solving the original optimization based setup of the problem where, a bilinear matrix equation with rank and norm constraint needs to be solved. Section 8.7 discusses the complexity and finite termination of the overall algorithm as well as the differences to the LP based algorithm in [Pol97]. The here computed weight W of the weighted vector norm is of small size and all the considered steps during the construction are elementary and scale well for large systems. Therefore, this algorithm is practically superior to the purely optimization based search, confer also the examples in Section 8.10.

In Sections 8.8 and 8.9 it is indicated how the result can be used in the aforementioned context of receding horizon control for piecewise affine systems in order to a priori guarantee closed-loop stability and feasibility for all time.

8.2 Software Implementation

The presented algorithm is implemented in the Multi-Parametric Toolbox (MPT) [KGB04, KGBC06, KGBM03] for MATLAB®. The toolbox can be downloaded free of charge at: http://control.ee.ethz.ch/~mpt/

8.3 Stability Theory for Linear Systems Using Linear Vector Norms

Recall that the vector ∞- and 1-norm [HJ85] of $z \in \mathbb{C}^{n_z}$ (or $\mathbb{R}^{n_z}$) is defined as

$$\|z\|_\infty := \max_{1 \le i \le n_z} |z_i| \quad \text{and} \quad \|z\|_1 := \sum_{i=1}^{n_z} |z_i|,$$

respectively, where z_i is the i-th element of z. As in [HJ85], lets denote the matrix ∞-norm (or *maximum row-sum norm*) and the matrix 1-norm (or *maximum column-sum norm*) of $S \in \mathbb{C}^{m \times n_z}$ (or $\mathbb{R}^{m \times n_z}$) by

$$\|S\|_\infty := \sup_{z \ne 0_{n_z}} \frac{\|Sz\|_\infty}{\|z\|_\infty} = \max_{1 \le i \le m} \sum_{j=1}^{n_z} |S_{i,j}| \quad \text{and}$$

$$\|S\|_1 := \sup_{z \ne 0_{n_z}} \frac{\|Sz\|_1}{\|z\|_1} = \max_{1 \le j \le n_z} \sum_{i=1}^{m} |S_{i,j}|,$$

respectively, where $S_{i,j}$ is the (i,j)-th element of S.

Here, we recapitulate the main existence theorem of the considered linear vector norm based Lyapunov functions for discrete-time systems proved in [KAS92, Pol95].

Theorem 8.1 (Lyapunov stability for discrete-time systems). Let $W \in \mathbb{R}^{m \times n_x}$ with $\text{rank}(W) = n_x$ and $p \in \{1, \infty\}$. The function

$$V(x) = \|Wx\|_p \tag{8.2}$$

is a Lyapunov function for the discrete-time system

$$x(t+1) = Ax(t), \quad t \ge 0, \tag{8.3}$$

with $x(t) \in \mathbb{R}^{n_x}$ and $A \in \mathbb{R}^{n_x \times n_x}$, *if and only if* there exists a matrix $Z \in \mathbb{R}^{m \times m}$, such that

$$WA = ZW, \tag{8.4a}$$

$$\|Z\|_p < 1. \tag{8.4b}$$
∎

Note that if $W \in \mathbb{C}^{m \times n_x}$ ($Z \in \mathbb{C}^{m \times m}$) then Theorem 8.1 is merely sufficient and not necessary, i.e. if a complex pair (W, Z) fulfills Equations (8.4a)–(8.4b) and the rank constraint on W then $\|Wx\|_p$ is a Lyapunov function for system (8.3), cf. [KAS92, Pol95] and the counter-proof example within.

A purely optimization based search for a real solution pair (W, Z) of the given bilinear matrix equation (8.4a) with rank constraint on W, norm constraint (8.4b), and a priori unknown m is in general difficult to solve and numerically demanding due to its nonlinear and non-convex character, cf. for example Section 8.10. To the author's knowledge no efficient, nor polynomial time algorithm is known for this problem.

In the following it is shown that the general problem can be solved very efficiently by a decomposition procedure using a particular structure for the pair (W, Z) and the solution to a finite and bounded sequence of simple feasibility LPs with few constraints or, equivalently, by utilizing simple algebraic tests.

Remark 8.2 (Vector 1-/∞-norm). Note that in the following a construction of a real W is mostly performed for the vector ∞-norm case ($p = \infty$). A completely equivalent construction for the vector 1-norm case ($p = 1$) can be performed, where the scaling matrices, denoted in the following with H, have to be similarly adjusted. This is indicated throughout the following text. □

In the following, the abovementioned overall problem is decomposed into several simpler problems. Throughout Section 8.4 and 8.5 these small 'building blocks' will be presented which eventually will lead to the fully implementable main Theorem 8.10 at the end of Section 8.5 that covers the Lyapunov function construction for the general case.

8.4 Complex W for Discrete-Time Systems

First, lets define the concept of eigenvalue multiplicity or eigenvalue multiplicity gap.

Definition 8.3 (Eigenvalue multiplicity). [LT85, HJ85] Assume the eigenvalue λ has the algebraic multiplicity l_{alg} and geometric multiplicity l_{geo}, then we say the eigenvalue λ has the *eigenvalue multiplicity (gap)* $l := l_{\text{alg}} - l_{\text{geo}} \geq 0$. □

Note that we often imprecisely say the 'eigenvalue λ has multiplicity l' when we actually mean it has algebraic multiplicity l_{alg}. An eigenvalue is called *simple* if $l_{\text{alg}} = l_{\text{geo}} = 1$.

For the case of $W \in \mathbb{C}^{m \times n_x}$, [Pol95] indicated the following lemma.

Lemma 8.4 (Jordan form). Let $p = \infty$ and A_{JF} be the *Jordan form* (JF) [LT85, HJ85] of $A \in \mathbb{R}^{n_x \times n_x}$ given by

$$A_{\text{JF}} := T_{\text{JF}}^{-1} A T_{\text{JF}} = \begin{bmatrix} \lambda_1 & r_1 & & 0 \\ & \ddots & \ddots & \\ & & \ddots & r_{n_x-1} \\ 0 & & & \lambda_{n_x} \end{bmatrix} \in \mathbb{C}^{n_x \times n_x},$$

where $\lambda_i \in \mathbb{C}$ is an eigenvalue of A with $|\lambda_i| < 1$, $r_i = 0$ if the geometrical and algebraic multiplicity [LT85, HJ85] of the eigenvalue λ_i are equal, and $r_i \neq 0$ (usually $r_i = 1$) otherwise. Let the scaling matrix $H_R := \text{diag}(h_1, \ldots, h_{n_x})$ with $h_{n_x} = 1$ and $h_i > 0$ such that $|r_i|h_i + (|\lambda_i| - 1)h_{i+1} < 0$ for $i = 1, \ldots, n_x - 1$.
Then the pair

$$Z_R := H_R A_{\text{JF}} H_R^{-1} \in \mathbb{C}^{n_x \times n_x},$$
$$W_R := H_R \in \mathbb{R}^{n_x \times n_x}$$

fulfills $\text{rank}(W_R) = n_x$ and Equations (8.4a)–(8.4b) for A_{JF}. $W := H_R T_{\text{JF}}^{-1} \in \mathbb{C}^{n_x \times n_x}$, $Z := Z_R$. □

Proof. Due to $h_i > 0$ follows $\text{rank}(W_R) = n_x$. $W_R A_{\text{JF}} = Z_R W_R \Leftrightarrow H_R A_{\text{JF}} = H_R A_{\text{JF}} H_R^{-1} H_R \Leftrightarrow H_R A_{\text{JF}} = H_R A_{\text{JF}}$. For brevity lets denote with $[Z_R]_i$ the i-th row of Z_R, then $[Z_R]_i = [0 \ldots 0, \lambda_i, \frac{h_i}{h_{i+1}} r_i, 0 \ldots 0]$ for $i = 1, \ldots, n_x - 1$ and $[Z_R]_{n_x} = [0 \ldots 0, \lambda_{n_x}]$. Thus $\|Z_R\|_\infty < 1$. ■

Note that if λ_i is an eigenvalue of A with equal geometric and algebraic multiplicity then one can simply chose $h_i = 1$. Moreover, if all $\lambda_i \in \mathbb{R}$ then A_{JF}, T_{JF}, and Z_R can be chosen to be real.

8.5 Construction of a Real W for Discrete-Time Systems

For control purposes, as for example in [BBM02, BCM06] or in Part II, it is usually desired to compute a W in the real vector space. One of the reasons is that if W is real and of finite dimension then the level sets of $V(x) = \|Wx\|_p$ are of polytopic nature and thus the function $V(\cdot)$ can be represented in a simple piecewise linear form. This, in turn, leads to purely polyhedral manipulations in further controller computations, see e.g. Part II. Furthermore, the sufficient and necessary property of Theorem 8.1 points to the choice of a real valued W.

If *all* the eigenvalues of A are real then one can simply use the formulation in Lemma 8.4 to compute the real W, as mentioned at the end of Section 8.4. However if one of the eigenvalues are in the complex domain, the construction of a real W becomes more involved, as shown in the following.

It is well known [LT85, HJ85] that with the following Lemma one can always transform the complex Jordan form of Lemma 8.4 into a particular real Jordan form.

Lemma 8.5 (Scaled real Jordan form). [LT85, HJ85] Assume the matrix $A \in \mathbb{R}^{2l \times 2l}$ has one complex eigenvalue pair $\{\lambda, \bar{\lambda}\} = \sigma \pm j\omega$ with eigenvalue multiplicity l, then the Jordan decomposition of A can be given by

$$A_{\text{JF}} := T_{\text{JF}}^{-1} A \, T_{\text{JF}} = \begin{bmatrix} J_l(\lambda) & 0 \\ 0 & J_l(\bar{\lambda}) \end{bmatrix} \quad \text{with} \quad J_l(\lambda) := \begin{bmatrix} \lambda & r & & 0 \\ & \ddots & \ddots & \\ & & \ddots & r \\ 0 & & & \lambda \end{bmatrix} \in \mathbb{C}^{l \times l},$$

and $r \neq 0$ (usually $r = 1$). Using a scaling matrix $H := \text{diag}(1, h^{-1}, h^{-2}, \ldots, h^{1-2l}) \in \mathbb{R}^{2l \times 2l}$ with $h \neq 0$ and

$$T_{\text{RJF}}^{(l)} := \frac{1}{\sqrt{2}} \begin{bmatrix} 1 & -j & 0 & \cdots & & 0 \\ & \ddots & \ddots & & & \\ 0 & \cdots & 0 & 1 & -j & \\ 1 & j & 0 & \cdots & & 0 \\ & \ddots & \ddots & & & \\ 0 & \cdots & 0 & 1 & & j \end{bmatrix} \in \mathbb{C}^{2l \times 2l}$$

one obtains the corresponding 'scaled' *real Jordan form* (RJF) as

$$A_{\mathbb{C}}^{(l)}(\lambda) := \tilde{T}_{\text{RJF}}^{-1} A \tilde{T}_{\text{RJF}} := (T_{\text{RJF}}^{(l)})^{\text{H}} H A_{\text{JF}} H^{-1} T_{\text{RJF}}^{(l)} = \begin{bmatrix} A_C & rh\,I_2 & & 0 \\ & \ddots & \ddots & \\ & & \ddots & rh\,I_2 \\ 0 & & & A_C \end{bmatrix},$$

where $A_{\mathbb{C}}^{(l)}(\lambda) \in \mathbb{R}^{2l \times 2l}$, the superscript H denotes the complex conjugate transpose, $A_C = \begin{bmatrix} \sigma & \omega \\ -\omega & \sigma \end{bmatrix}$, $I_2 = \begin{bmatrix} 1 & 0 \\ 0 & 1 \end{bmatrix}$, and $\tilde{T}_{\text{RJF}} \in \mathbb{R}^{2l \times 2l}$. ∎

One of the main new core contributions for computing a Lyapunov function $V(x) = \|Wx\|_p$ for an arbitrary stable A is based on the following lemma.

Lemma 8.6 (Simple complex eigenvalues). The function $V(x) = \|W_m x\|_p$, $p \in \{1, \infty\}$ and $m \in \mathbb{N}_{\geq 2}$, with the given real $(m \times 2)$-matrix

$$W_m := \begin{bmatrix} \vdots & \vdots \\ \cos\left(\frac{(i-1)\pi}{m}\right) & \sin\left(\frac{(i-1)\pi}{m}\right) \\ \vdots & \vdots \end{bmatrix}, \quad i = 1, \ldots, m, \qquad (8.6)$$

8.5 Construction of a Real W for Discrete-Time Systems

is a Lyapunov function for the stable discrete-time system

$$x(t+1) = A_C x(t), \quad t \geq 0, \tag{8.7}$$

with $A_C = \begin{bmatrix} \sigma & \omega \\ -\omega & \sigma \end{bmatrix}$, and $\sigma, \omega \in \mathbb{R}$, if the relative interior of the polytope

$$\mathcal{Z} = \{ z \in \mathbb{R}^m \mid [\sigma, \omega]' = W_m' z \text{ and } \|z\|_1 \leq 1 \} \tag{8.8}$$

is non-empty. □

Proof. Recall that $[X]_i$ denotes i-th row of some matrix X. $c_i := \cos(\frac{(i-1)\pi}{m})$, $s_i := \sin(\frac{(i-1)\pi}{m})$, e_i is the i-th column of the $(m \times m)$-unity-matrix I_m. Take the following skew-symmetric circular $(m \times m)$-matrix

$$Z_m := \begin{bmatrix} z_1 & z_2 & z_3 & \cdots & z_m \\ -z_m & z_1 & z_2 & \cdots & z_{m-1} \\ & & \ddots & \ddots & \\ & & & \ddots & z_2 \\ -z_2 & -z_3 & -z_4 & \cdots & z_1 \end{bmatrix}$$

with $z_i \in \mathbb{R}$, $i = 1, \ldots, m$. Then the following relationships hold

$$[W_m]_i A_C = \ldots = [\sigma, \omega] T_{cs}(i), \quad T_{cs}(i) := \begin{bmatrix} c_i & s_i \\ -s_i & c_i \end{bmatrix},$$

$$[Z_m]_i = z' E_m(i), \quad E_m(i) := [-e_{m-i+2}, \ldots, -e_m, e_1, \ldots, e_{m-i+1}],$$

where $E_m(i) \in \mathbb{R}^{m \times m}$ and $z := [z_1, \ldots, z_m]'$. It follows from the i-th row of $W_m A_C = Z_m W_m$

$$[W_m]_i A_C = [Z_m]_i W_m \Leftrightarrow [\sigma, \omega] T_{cs}(i) = z' E_m(i) W_m$$
$$\Leftrightarrow [\sigma, \omega] = z' E_m(i) W_m T_{cs}'(i) = \ldots = z' W_m, \tag{8.10}$$

where $i = 1, \ldots, m$. Note that the relation (8.10) between σ, ω, and z is independent of i and thus each row of $W_m A_C = Z_m W_m$ is equivalent. Moreover, due to the structure of Z_m it follows immediately that for $p \in \{1, \infty\}$ it holds $\|Z_m\|_p < 1 \Leftrightarrow \|z\|_1 < 1$. rank$(W_m) = 2 = n_x$. Theorem 8.1 completes the proof. ∎

Lemma 8.6 states a test if $\|W_m x\|_p$ is a Lyapunov function for the given system (8.7). This test can performed by solving a *simple feasibility LP* with domain relint$(\mathcal{Z})$ in (8.8). As an example Figure 8.1 depicts an ∞-norm based Lyapunov function $\|W_m x\|_\infty$ with $m = 17$ for a two-dimensional oscillating discrete-time system with an eigenvalue pair close to the stability boundary, i.e. the spectral radius $\rho(A) = |\lambda| = 1 - 10^{-5}$.

It follows from Lemma 8.6 for which set of systems or eigenvalue distributions the same Lyapunov function is valid, i.e.

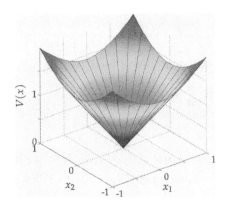
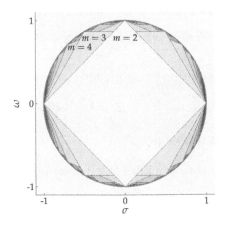

Fig. 8.1: Linear vector norm based Lyapunov function for an 'almost' unstable oscillating two-dimensional system.

Fig. 8.2: Parameter space of $[\sigma, \omega]'$ in dependence of m for which $\|W_m x\|_{\{1,\infty\}}$ is a Lyapunov function for the discrete-time system (8.11).

Corollary 8.7 (Parameter space of σ and ω). For a fixed $m \in \mathbb{N}_{\geq 2}$, $V(x) = \|W_m x\|_p$ with $p \in \{1, \infty\}$ as defined in Lemma 8.6 is a Lyapunov function for *all* systems

$$x(t+1) = \begin{bmatrix} \sigma & \omega \\ -\omega & \sigma \end{bmatrix} x(t), \quad t \geq 0, \tag{8.11}$$

that fulfill

$$\begin{bmatrix} \sigma \\ \omega \end{bmatrix} \in \mathrm{int}(\mathcal{P}_m), \tag{8.12a}$$

where

$$\mathcal{P}_m := \mathrm{conv}\left(\left\{\pm \begin{bmatrix} \cos\left(\frac{(i-1)\pi}{m}\right) \\ \sin\left(\frac{(i-1)\pi}{m}\right) \end{bmatrix}\right\}_{i=1}^m\right) \tag{8.12b}$$

and $\mathrm{conv}(\cdot)$ is the convex hull, i.e. for $V(x)$ to be a Lyapunov function for system (8.11), $[\sigma, \omega]'$ has to lie in the interior of the regular polygon with $2m$ vertices with one vertex at $[1, 0]'$ inscribed in the unit circle, cf. Figure 8.2. □

Proof. Consider Lemma 8.6. The closure of $\|z\|_1 < 1$ is a convex polytope in $\mathbb{R}^m$ with vertices at $z = \pm e_i$, $i = 1, \ldots, m$. $[\sigma, \omega]' = W_m' z$ is a linear projection onto the (σ, ω)-space. ∎

We want to point out that the search for the proper m for continuous-time systems assuming the structure of W_m, as it is presented in [Pol95], is simpler than in the discrete-time case. This is due to the fact that as one increases

8.5 CONSTRUCTION OF A REAL W FOR DISCRETE-TIME SYSTEMS

m one obtains supersets in the (σ, ω)-parameter space $\tilde{\mathcal{P}}_m$, i.e. $\tilde{\mathcal{P}}_m \subset \tilde{\mathcal{P}}_{m+1}$, indicating the feasibility of the problem and thus if there exists a W_{m^*} for a particular m^* then there also exists a feasible W_m for all $m \geq m^*$. This is not true for the discrete-time case (using the particular formulation for W_m), cf. Figure 8.2. In addition, note that the simpler structure of Z_m as proposed in [Pol95], where $z_i = 0$ for $i = 3, \ldots, m$, is *not* sufficient for the class of discrete-time systems, because it would correspond only to the parameter space $\tilde{\mathcal{P}}_m :=$ conv$(\{\pm\begin{bmatrix}1\\0\end{bmatrix}, \pm\begin{bmatrix}\cos(\pi/m)\\\sin(\pi/m)\end{bmatrix}\})$, which for any $m \in \mathbb{N}_{\geq 2}$ misses large parts of the unit circle. Thus for a large set of stable systems the proposed formulation would not lead to a successful computation of W_m, respectively W. Take for example the system with parameters $[\sigma, \omega]' = [-\frac{1}{2}, \frac{1}{2} + \varepsilon]'$, $\varepsilon \in (0, \frac{\sqrt{3}-1}{2})$, a Lyapunov function of the considered type does not exist using a Z_m as proposed in [Pol95].

From Corollary 8.7 one can easily see the direct relation to the commonly[1] used vector 2-norm based Lyapunov function as m tends to infinity.

Corollary 8.8 (Relation to $\|x\|_2$). $\tilde{V}(x) = \|x\|_2$ is a Lyapunov function for *all* discrete-time systems

$$x(t+1) = \begin{bmatrix} \sigma & \omega \\ -\omega & \sigma \end{bmatrix} x(t), \quad t \geq 0,$$

that fulfill $\|[\sigma, \omega]'\|_2 < 1$, i.e. are stable. □

Proof. It holds for all $m \in \mathbb{N}_{\geq 2}$ that $r_m\|x\|_2 \leq \|W_m x\|_\infty \leq \|x\|_2$, where $r_m = \frac{1}{2}\|[1 + \cos(\frac{\pi}{m}), \sin(\frac{\pi}{m})]'\|_2$. Moreover, r_m is monotonically increasing with m and $\lim_{m\to\infty} r_m = 1$. Thus, $\lim_{m\to\infty} \|W_m x\|_\infty = \|x\|_2$. Similarly it holds for the case $p = 1$. Now it is left to show that, as m increases, $\mathcal{P}_m$ converges to the complete interior of the unit circle: by construction $[c_i, s_i]'$ lies on the unit circle for any $m, i \in \mathbb{N}$ and $\mathcal{P}_m$ inscribes a convex set inside the unit circle. The limit of the area of the regular polygon $\mathcal{P}_m$ is given by $\lim_{m\to\infty} m \sin(\pi/m) = \pi$. ∎

Of course it is easy to test if $\|x\|_2$ is a Lyapunov function for the considered simple system, but here the aim was to show the direct relation to the linear vector norm based Lyapunov functions considered.

In the case of the existence of complex eigenvalues with eigenvalue multiplicity, one needs to modify Lemma 8.6 with a particular scaling and augmentation procedure:

Lemma 8.9 (Complex eigenvalues with multiplicity). Let the matrix $A \subset \mathbb{R}^{2l \times 2l}$ have one complex eigenvalue pair $\{\lambda, \bar{\lambda}\} = \sigma \pm j\omega$ with eigenvalue multiplicity l, then the real Jordan form, cf. Lemma 8.5, of A can be given by

---

[1] In control the most common choice for Lyapunov functions are of the quadratic form $x'\tilde{W}x$. Note, that it is immediate that if $\|Wx\|_2$ is a Lyapunov function then $\|Wx\|_2^2 = x'W'Wx$ is one.

$$A_C^{(l)}(\lambda) := \tilde{T}_{RJF}^{-1} A \tilde{T}_{RJF} = \begin{bmatrix} A_C & \tilde{r} I_2 & & 0 \\ & \ddots & \ddots & \\ & & \ddots & \tilde{r} I_2 \\ 0 & & & A_C \end{bmatrix} \in \mathbb{R}^{2l \times 2l},$$

with $A_C = \begin{bmatrix} \sigma & \omega \\ -\omega & \sigma \end{bmatrix}$, $I_2 = \begin{bmatrix} 1 & 0 \\ 0 & 1 \end{bmatrix}$, and the pair (W_m, Z_m), as in Lemma 8.6, solves

$$W_m A_C = Z_m W_m, \qquad \|Z_m\|_p < 1 - \tilde{r}$$

with $p \in \{1, \infty\}$ and

$$0 \leq \tilde{r} < 1 - \sigma^2 - \omega^2.$$

Then, using an appropriate H to scale $\tilde{r}$ (cf. Lemma 8.5), the pair

$$W_m^{(l)} := \begin{bmatrix} W_m & & 0 \\ & \ddots & \\ 0 & & W_m \end{bmatrix} \in \mathbb{R}^{lm \times 2l}, \quad Z_m^{(l)} := \begin{bmatrix} Z_m & \tilde{r} I_m & & 0 \\ & \ddots & \ddots & \\ & & \ddots & \tilde{r} I_m \\ 0 & & & Z_m \end{bmatrix} \in \mathbb{R}^{lm \times lm}$$

solves

$$W_m^{(l)} A_C^{(l)}(\lambda) = Z_m^{(l)} W_m^{(l)}, \qquad \|Z_m^{(l)}\|_p < 1, \qquad \text{and} \quad \text{rank}\left(W_m^{(l)}\right) = 2l. \qquad \square$$

Proof. $W_m^{(l)} A_C^{(l)}(\lambda) = Z_m^{(l)} W_m^{(l)} \Leftrightarrow \ldots \Leftrightarrow W_m A_C = Z_m W_m$. Due to the structure of Z_m it follows $\|Z_m^{(l)}\|_\infty = \|Z_m^{(l)}\|_1 = \max\{\|Z_m\|_\infty, \|Z_m + \tilde{r} I_m\|_\infty\} \leq \max\{\|Z_m\|_\infty, \|Z_m\|_\infty + \tilde{r}\} < \max\{1 - \tilde{r}, 1\} = 1$. $\text{rank}(W_m^{(l)}) = l \, \text{rank}(W_m) = 2l$. ∎

The following theorem combines the above mentioned elementary ideas to compute a linear vector norm based Lyapunov function for *general stable linear discrete-time systems*.

Theorem 8.10. Let the stable discrete-time system be given by

$$x(t+1) = Ax(t), \quad t \geq 0,$$

where $x(t) \in \mathbb{R}^{n_x}$, $A \in \mathbb{R}^{n_x \times n_x}$, and $T \in \mathbb{R}^{n_x \times n_x}$ transforms A into the real Jordan form

$$A_{RJF} := T^{-1} A T = \begin{bmatrix} A_R & & & 0 \\ & A_C^{(l_1)}(\lambda_{C,1}) & & \\ & & \ddots & \\ 0 & & & A_C^{(l_{n_C})}(\lambda_{C,n_C}) \end{bmatrix} \in \mathbb{R}^{n_x \times n_x}, \quad (8.15a)$$

8.5 Construction of a Real W for Discrete-Time Systems

where

$$A_R = \begin{bmatrix} \lambda_{R,1} & r_1 & & 0 \\ & \ddots & \ddots & \\ & & \ddots & r_{n_R-1} \\ 0 & & & \lambda_{R,n_R} \end{bmatrix} \in \mathbb{R}^{n_R \times n_R}, \quad \text{and} \quad (8.15b)$$

$$A_C^{(l_i)}(\lambda_{C,i}) = \begin{bmatrix} \begin{bmatrix} \sigma_i & \omega_i \\ -\omega_i & \sigma_i \end{bmatrix} & r_i I_2 & & 0 \\ & \ddots & \ddots & \\ & & \ddots & r_i I_2 \\ 0 & & & \begin{bmatrix} \sigma_i & \omega_i \\ -\omega_i & \sigma_i \end{bmatrix} \end{bmatrix} \in \mathbb{R}^{2l_i \times 2l_i}. \quad (8.15c)$$

Here $\lambda_{R,i}$ corresponds to the i-th real eigenvalue of A, $\{\lambda_{C,i}, \bar{\lambda}_{C,i}\} = \sigma_i \pm j\omega_i$ with $\sigma_i, \omega_i \in \mathbb{R}$, corresponds to the i-th complex eigenvalue pair of A with eigenvalue multiplicity l_i. $r_i = 0$ if the geometrical and algebraic multiplicity of the eigenvalues $\lambda_{R,i}$ (resp. $\lambda_{C,i}$) are equal, and $r_i \neq 0$ (usually $r_i = 1$) otherwise.

Then there exists a Lyapunov function $V(x) = \|Wx\|_p$, $p \in \{1, \infty\}$, for the system with $W = \tilde{W}T^{-1}$, where

$$\tilde{W} := \begin{bmatrix} W_R & & & 0 \\ & W_{m_1}^{(l_1)} & & \\ & & \ddots & \\ 0 & & & W_{m_{n_C}}^{(l_{n_C})} \end{bmatrix} \in \mathbb{R}^{\tilde{m} \times (n_R + 2n_C)} \quad (8.16)$$

with $\tilde{m} = n_R + \sum_{i=1}^{n_C} m_i$ and $W_R, W_{m_i}^{(l_i)}, i = 1, \ldots, n_C$, as defined and computed in Lemma 8.4–8.9. □

Proof. Take

$$Z = \begin{bmatrix} Z_R & & & 0 \\ & Z_{m_1}^{(l_1)} & & \\ & & \ddots & \\ 0 & & & Z_{m_{n_C}}^{(l_{n_C})} \end{bmatrix} \in \mathbb{R}^{\tilde{m} \times \tilde{m}}.$$

Then $WA = ZW \Leftrightarrow \tilde{W}T^{-1}AT = Z\tilde{W} \Leftrightarrow \tilde{W}A_{RJF} = Z\tilde{W}$, which is equivalent to (1.) $W_R A_R = Z_R W_R$, cf. Lemma 8.4, and (2.) $W_{m_i}^{(l_i)} A_C^{(l_i)}(\lambda_{C,i}) = Z_{m_i}^{(l_i)} W_{m_i}^{(l_i)}$, $i = 1, \ldots, n_C$, cf. Lemma 8.9. $\text{rank}(W) = \text{rank}(\tilde{W}) = n_R + \sum_{i=1}^{n_C} 2l_i = n_x$. For $p \in \{1, \infty\}$ it holds $\|Z\|_p = \max_i \left\{ \|Z_R\|_p, \|Z_{m_i}^{(l_i)}\|_p \right\} < 1$. Theorem 8.1 completes the proof. ∎

It should be noted that Theorem 8.10 is general and not, as in [KAS92, Bit88], restricted to eigenvalues of A which lie in the convex set spanned by the vertices $1, +j, -1, -j$ in the complex plane.

8.6 Simple Algebraic Test to Substitute the Feasibility LP (8.8)

Lemma 8.6 implies that one needs to test if for a fixed m the polytope $\mathcal{Z}$ in (8.8) has a non-empty relative interior. If this is not the case one needs to increase m and perform a similar test again. From Corollary 8.7 we have that the projection of $\mathcal{Z}$ onto the (σ, ω)-space form regular polygons which, as m increases, approximate the unit circle arbitrarily well, cf. Figure 8.2–8.4.

Therefore, the equivalent geometric test is: in which polygon (or in which sector of which polygon) does the point $[\sigma_i, \omega_i]'$ lie. How to derive and perform this test is explained in the following.

Given the simple complex eigenvalue pair

$$\{\lambda_i, \bar{\lambda}_i\} = \sigma_i \pm j \omega_i = r_i\, e^{\pm j \varphi_i},$$

where r_i is the modulus and φ_i is the angle of λ_i, respectively. λ_i corresponds to the point $L_i := [\sigma_i, \omega_i]'$ in the (σ, ω)-space. Due to the symmetric shape of the regular polygons $\mathcal{P}_m$ one needs only to consider the first orthant. Thus, yields

$$r_i = \sqrt{\sigma_i^2 + \omega_i^2} \quad \text{and} \quad \varphi_i = \begin{cases} \frac{\pi}{2}, & \text{if } \sigma_i = 0, \\ \arctan\left(\frac{|\omega_i|}{|\sigma_i|}\right), & \text{if } \sigma_i \neq 0. \end{cases}$$

The vertices of the polygon $\mathcal{P}_m$ nearest to L_i are

$$\underline{L}_m := \begin{bmatrix} \cos \underline{\varphi} \\ \sin \underline{\varphi} \end{bmatrix} \quad \text{and} \quad \bar{L}_m := \begin{bmatrix} \cos \bar{\varphi} \\ \sin \bar{\varphi} \end{bmatrix},$$

where $\underline{\varphi} = \lfloor \frac{\varphi_i m}{\pi} \rfloor \cdot \frac{\pi}{m}$ and $\bar{\varphi} = (\lfloor \frac{\varphi_i m}{\pi} \rfloor + 1) \cdot \frac{\pi}{m}$. The function $\lfloor \cdot \rfloor$ denotes rounding to the nearest smaller integer. Thus, $0 \leq \underline{\varphi} \leq \varphi_i \leq \bar{\varphi} \leq \pi$ and $\underline{\varphi} < \bar{\varphi}$. The sector of interest is spanned by the origin, $\underline{L}_m$, and $\bar{L}_m$. Let X be the point on the connecting line between $\underline{L}_m$ and $\bar{L}_m$ having the same angle as L_i, i.e.

$$X: \quad \gamma \underline{L}_m + (1 - \gamma) \bar{L}_m = r_X \begin{bmatrix} \cos \varphi_i \\ \sin \varphi_i \end{bmatrix} \quad (8.17)$$

with $\gamma \in [0, 1]$ and $r_X < 1$. Then the only test that is left is if $r_i < r_X$. Solving (8.17) for r_X, and using some geometry, the test results in: iterate on m and stop if m fulfills

$$r_i < r_X = \frac{\cos\left(\frac{\pi}{2m}\right)}{\cos\left(\varphi_i - \frac{\pi}{2m}\left(2 \lfloor \frac{\varphi_i m}{\pi} \rfloor + 1\right)\right)}.$$

8.7 Computational Complexity and Finite Termination

As mentioned above, a purely optimization based search for a solution pair $(W, Z) \in \mathbb{R}^{m \times n_x} \times \mathbb{R}^{m \times m}$ to (8.4a)–(8.4b), for a priori unknown m, is in general very difficult due to the nonlinear and non-convex nature of the problem, cf. also Section 8.10 for an example. To the author's knowledge no efficient, nor polynomial time algorithm is known for this problem.

In comparison, for the proposed approach the complexity of the computation of a Lyapunov function $V(x) = \|Wx\|_{\{1,\infty\}}$ is basically driven by two factors: *first*, the Jordan decomposition (8.15a)–(8.15c) and, *second*, the computation of a finite and bounded sequence of feasibility LPs with the simple constraint (8.8), cf. Lemma 8.6. Moreover, it is shown in Section 8.6 that, instead of solving the feasibility LP, one can equivalently perform a simple algebraic test in order to obtain the minimal m for $\|W_m x\|_{\{1,\infty\}}$.

Due to the second part being a finite and bounded number of simple operations, the possible bottleneck of the algorithm lies mainly in the decomposition of the system matrix A. It should be mentioned that the Jordan decomposition might be, depending on the conditioning and structure of A, numerically problematic. However, this does not necessarily imply that generally the Jordan decomposition is numerically (more) complex in higher dimensional spaces. Even more so, in most practical and 'almost' all randomly chosen problems the Jordan decomposition simplifies to a diagonal eigenvalue decomposition, because of the non-robustness or sensitivity[2] of the Jordan form itself. So for control problems of practical size, i.e. for example systems with dimension less than 2000, is can be assumed that this is not of significant relevance. Confer also the example Section 8.10.

8.7.1 Boundedness of m and Finite Termination

A rather conservative upper bound $m_{\max}(\lambda_{C,i})$ on the minimal $m(\lambda_{C,i})$ for the complex eigenvalue pair $\{\lambda_{C,i}, \overline{\lambda}_{C,i}\} = \sigma_i \pm j\omega_i$ with $\omega_i \neq 0$ can be found via the inner radius of regular polygons $\mathcal{P}_m$; thus we obtain

$$2 \leq m(\lambda_{C,i}) \leq m_{\max}(\lambda_{C,i}) := \left\lceil \frac{\pi}{2 \arccos\left(\sqrt{\sigma_i^2 + \omega_i^2}\right)} \right\rceil < \infty,$$

---

[2] An arbitrarily small generic perturbation to the system matrix A leads to a significant change in the eigenvectors and eigenvalues of the matrix [GW94].

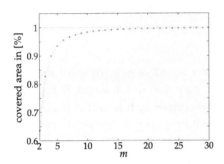
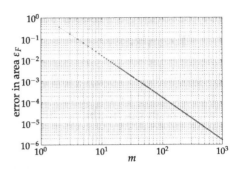

Fig. 8.3: Covered area of the (σ, ω)-parameter space with respect to the unit circle for which $\|W_m x\|_{\{1,\infty\}}$ is a Lyapunov function.

Fig. 8.4: Error $\varepsilon_F(m) = 1 - \frac{F(m)}{\pi}$ of the covered area $F(m)$ of the (σ, ω)-parameter space with respect to the unit circle for which $\|W_m x\|_{\{1,\infty\}}$ is a Lyapunov function.

where the function $\lceil \cdot \rceil$ denotes rounding to the nearest larger integer. It follows with $\tilde{m}$ in (8.16) for the general case that the overall

$$m \leq n_R + \sum_{i=1}^{n_C} l_i \cdot m_{\max}(\lambda_{C,i}) < \infty,$$

where n_R is the total number of real eigenvalues, n_C is the number of complex eigenvalues, and l_i is the eigenvalue multiplicity of the complex eigenvalue pair $\{\lambda_{C,i}, \overline{\lambda}_{C,i}\}$. Hence, the guaranteed successful finite termination of the construction of W.

We point out that in this framework one needs to check the stability of the linear system, which is a trivial task, before starting to compute a Lyapunov function of the considered type. In the case of an unstable system the proposed algorithm would continue to iterate. However, the same would happen for the algorithms proposed in [Pol97, Pol95].

To indicate the 'practical' complexity of the problem, one obtains from Corollary 8.7 that $\|W_m x\|_{\{1,\infty\}}$ with $m \leq 13$ already resembles a Lyapunov function for more than 99% percent of all stable oscillating systems of the type (8.11), cf. Figure 8.3. However, if A has a spectral radius close to 1, in the worst case one would need a large (but bounded) number m to approximate the unit circle appropriately, cf. Figure 8.4. The worst case, however, can only appear if the angle φ_i of the eigenvalue $\lambda_i = \sigma_i + j\omega_i = |\lambda_i| e^{j\varphi_i}$ is an irrational multiple of π, i.e. $\varphi_i = c\pi$ with $c \notin \mathbb{Z}$. Figure 8.5 depicts such a worst case for a two-dimensional system, where the spectral radius is $\rho(A) = 1 - 10^{-5}$ and $\varphi_i = \sqrt{2}\pi$. Even though this is considered to be the 'worst case', W consists only of 70 rows, i.e. $\|Wx\|_\infty$ with $W \in \mathbb{R}^{70 \times 2}$.

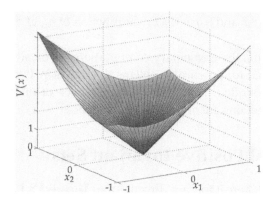

Fig. 8.5: Linear vector norm based Lyapunov function for an 'almost' unstable, highly oscillating, two-dimensional system with 'worst case' positioning of its eigenvalues.

8.7.2 A Complexity Comparison

As indicated above, the first result to constructively proof the existence of ∞-norm based Lyapunov functions for linear continuous-time systems was presented in [Pol97]. In contrast to the decomposition based approach of [Pol95] or the algorithm presented here, the algorithm in [Pol97] mainly relies on the use of a sequence of linear programs and scaling operations without explicity computing the eigenvalues of the system matrix A. A translation of the algorithm to the discrete-time case for ∞-norm based Lyapunov functions is (more or less) straightforward.

In [Pol97] one needs to provide an 'arbitrary' $W = W_0 \in \mathbb{R}^{m \times n_x}$ as well as m and then test if this particular W implies a Lyapunov function or not. This test is performed using m linear programs. If W_0 does not indicate a Lyapunov function, an additional scaling operation on W_0 is performed. In the case that this was not successful either, a different W_0 and/or increase of m is chosen and the same test is repeated.

However, the success of this technique is highly dependent on the choice of these two parameters. A guarantee of convergence (let alone finite termination) of this algorithm is not given. For example choosing W_0 'arbitrarily' (as stated in [Pol97]) from an uniform or normal distribution m tends to be much larger than the m needed in the algorithm presented here. Moreover, m varies significantly with the seed/run of the distribution chosen. When choosing W_0 as the approximation of the n_x-dimensional unit-sphere (as also suggested in [Pol97]) the algorithm seems to succeed with the same m as needed in the here presented algorithm. A proof or indication for finite termination is, however, lacking. In addition, as the system dimension n_x increases, this choice of W_0 will lead to an exponential number of rows m; whereas in the algorithm presented in Section 8.5 and 8.6 m is a linear function of n_x, cf. Section 8.7.1 and $\widetilde{m}$ in (8.16).

Thus it follows from the elementary computations involved in the here presented algorithm, that for practical problems the size of W is rather small and guaranteed to be bounded. And therefore the choice of linear vector norm based Lyapunov functions are of practical importance.

8.8 Polyhedral Positive Invariant Sets

Analogous to [Bit88], here it follows directly from Theorem 8.1 and the fact that level sets of Lyapunov functions are positively invariant [Vid93, Hah67] that the following holds.

Theorem 8.11 (Polyhedral positively invariant sets). The discrete-time system (8.3) is asymptotically stable if and only if there exist *polyhedral* positively invariant sets

$$\mathcal{O}(\nu) = \{x \in \mathbb{R}^{n_x} \mid \|Wx\|_p \leq \nu,\ WA = ZW,\ \|Z\|_p < 1, \operatorname{rank}(W) = n_x\},$$

where $\nu > 0$ and $p \in \{1, \infty\}$. ∎

In contrast to [Bit88], Theorem 8.11 is completely general and not restricted to the case when A has eigenvalues lying in the rhombus spanned by the vertices $1, +j, -1, -j$ in the complex plane.

We want to note, that in the context of receding horizon control, cf. Chapter 3 and 6 or [BBM02, MRRS00, Mac02, LHWB06], in order to a priori guarantee closed-loop stability and feasibility for all time, $\mathcal{O}(\nu)$ can e.g. serve as an invariant terminal target set $\mathcal{X}^f$. It is usually desired to maximize the target set for a fixed W, i.e. maximizing ν, within the pre-specified state constraints $\mathbb{X}$. This can be performed using a finite number of simple LPs if the set $\mathbb{X}$ is polytopic.

8.9 Guaranteeing Exponential Stability in Receding Horizon Control for Piecewise Affine Systems

Choosing a Lyapunov function according to Theorem 8.1 (conservatively) guarantees a Lyapunov decay of $\Delta V(x) < 0$ for all $x \neq \mathbb{0}_{n_x}$, where

$$\Delta V(x(t)) := V(x(t+1)) - V(x(t)) = \|WAx(t)\|_p - \|Wx(t)\|_p.$$

However, in receding horizon control it is needed to over-bound the Lyapunov decay by

$$\Delta V(x) \leq -\|\widetilde{P}x\|_p, \quad \forall x \neq \mathbb{0}_{n_x}, \tag{8.18}$$

8.9 Guaranteeing Exponential Stability in RHC for PWA Systems

in order to (conservatively) guarantee a priori closed-loop stability and feasibility for all time, cf. Section 3.4. Here $\widetilde{P}$ indicates the desired minimum decay, which is usually induced by the receding horizon stage cost $\ell(\,\cdot\,,\,\cdot\,)$. In the case of $p \in \{1, \infty\}$ it can be obtained as described in step 3 of Algorithm 8.12 or as on page 85.

The fulfillment of inequality (8.18) is easily achieved by scaling the aforementioned Lyapunov function $V(\,\cdot\,)$ with a scalar α_W, i.e. $V_{\text{new}}(x) := \alpha_W \|Wx\|_p$, where the scaling parameter is bounded by

$$\alpha_W \geq \frac{\|\widetilde{P}(W'W)^{-1}W'\|_p}{1 - \|Z\|_p}, \qquad (8.19)$$

cf. [BBM02, Prop. 1] and [Bor03].

Another possibility in order to reduce conservatism in α_W and to fulfill (8.18) is to minimize $\|Z\|_p$ in the algorithm presented in Section 8.4 and 8.5 by 'tweaking' the scaling matrices H. This can be performed with simple LPs. The success of this later technique is, however, highly dependent on the location of the eigenvalues of A.

Having said that, together with Section 8.8, one can now state the simple Algorithm 8.12 to compute the final penalty matrix P and terminal target set $\mathcal{X}^f$ which will a priori guarantee closed-loop *(local) exponential stability* and feasibility for all time of the on-line or off-line receding horizon control scheme for piecewise affine systems[3] with a linear cost function as it is considered in Chapter 6 (and Chapter 3). It should to be noted, however, that here the equilibrium point $\mathbb{0}_{n_x}$ needs to have an arbitrarily small neighborhood $\mathcal{N}(\mathbb{0}_{n_x})$ entirely in one of the dynamics $\mathcal{D}_d$ of the piecewise affine system (4.1). Thus there exists an $\mathcal{N}(\mathbb{0}_{n_x}) \neq \emptyset$, $d^\star \in \{1, \ldots, N_\mathcal{D}\}$, and a linear local stabilizing controller $\mu_{\text{loc}}(x) = Kx$ such that

$$[x'\ (Kx)']' \in \mathcal{D}_{d^\star} \qquad \text{for all} \quad x \in \mathcal{N}(\mathbb{0}_{n_x}).$$

Note that in most engineering applications this is not a restricting requirement.

Algorithm 8.12 (Exponentially stable closed-loop RHC)

1. design a local linear stabilizing controller $\mu_{\text{loc}}(x) = Kx$ around $x = \mathbb{0}_{n_x}$
2. compute W according to Theorem 8.10 for $A_{\text{loc}}^{\text{CL}} := A_{d^\star} + B_{d^\star}K$
3. compute α_W as in (8.19), $P := \alpha_W W$
   if $p = 1$, $\widetilde{P} := \begin{bmatrix} Q \\ RK \end{bmatrix}$
   if $p = \infty$, $\|\widetilde{P}x\|_\infty := \|Qx\|_\infty + \|RKx\|_\infty$ (computation via one mp-LP)

---
[3] Note that Algorithm 8.12 is equally applicable to *general nonlinear* systems (guaranteeing an exponentially stabilizing receding horizon controller) that exhibit local linear behavior, i.e. $x(t+1) = f(x(t), u(t)) = Ax(t) + Bu(t)$ for all $(x(t), u(t)) \in \mathcal{N}(\mathbb{0}_{n_x}) \times \mathbb{U}$.

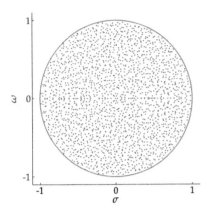

Fig. 8.6: Eigenvalue distribution of a 2000D stable discrete-time system.

Tab. 8.1: Example computation times of ∞-norm based Lyapunov functions using MATLAB®.

system dim. n_x	spectral radius $\rho(A)$	average CPU time
10	$1 - 10^{-3}$	$5.11 \cdot 10^{-3}$ sec
	$1 - 10^{-4}$	$1.18 \cdot 10^{-2}$ sec
	$1 - 10^{-5}$	$1.13 \cdot 10^{-1}$ sec
100	$1 - 10^{-3}$	$9.21 \cdot 10^{-2}$ sec
	$1 - 10^{-4}$	$1.10 \cdot 10^{-1}$ sec
	$1 - 10^{-5}$	$1.45 \cdot 10^{-1}$ sec
500	$1 - 10^{-3}$	4.02 sec
	$1 - 10^{-4}$	4.35 sec
	$1 - 10^{-5}$	4.52 sec
2000	$1 - 10^{-4}$	211.61 sec

4. compute the maximal (or some) level-set $\mathcal{O}(\nu_{\max})$ in Theorem 8.11 such that $\mathcal{O}(\nu_{\max})$ is a subset of $\mathcal{D}_{d^*}$, $\mathcal{X}^f := \mathcal{O}(\nu_{\max})$

8.10 Examples

Figure 8.6 depicts the eigenvalue distribution of a (almost) random 2000 dimensional stable discrete-time system. A subset of eigenvalues have high eigenvalue multiplicity and/or are close to the stability boundary, i.e. the spectral radius is $\rho(A) = 1 - 10^{-4}$. A Lyapunov function $V(x) = \|Wx\|_\infty$ for the system was found with $W \in \mathbb{R}^{2676 \times 2000}$ in 211.61 sec using a Pentium M, 1.6 GHz (1GB RAM) machine running MATLAB® 7.0.4 SP2, the LP solver of NAG® [Num02], and the algebraic test described in Section 8.6.

Table 8.1 indicates the average CPU time needed to compute a Lyapunov function on the aforementioned machine and setup. For each case the CPU time was averaged over 100 random systems. (Except for the 2000D case.) The system dimension or the location of the eigenvalues of the system do not seem to limit the computability of a Lyapunov function of the considered class using the proposed algorithm.

In contrast, an optimization based search in order to find a (W, Z) satisfying Equation (8.4a)–(8.4b), including the rank constraint on W, was not successful for system dimensions of order higher than 5 and not solvable in a reasonable amount of time (using available solvers) for system dimensions of order higher than 3, depending on the eigenvalue location. For example, an ∞-norm based Lyapunov function for a random 3-dimensional stable system was found in

58 sec using the internal global solver for bilinear problems of the optimization interface YALMIP [Löf04] together with the BMI solver PENBMI [PEN] by PENOPT as the local node solver. The here proposed approach found a Lyapunov function in $1.3 \cdot 10^{-3}$ sec for the very same system.

9
Stability Analysis

For a linear performance index the solution to the CFTOC problem is a time-varying piecewise affine function of the state. However, when a receding horizon control strategy is used, stability and/or feasibility for all time of the closed-loop system is not guaranteed a priori. In this chapter an algorithm is presented that by analyzing the CFTOC solution a posteriori extracts regions of the state space for which closed-loop stability and feasibility for all time can be guaranteed. The algorithm computes the maximal positively invariant set and stability region of a piecewise affine system by combining reachability analysis with some basic polyhedral manipulation. The simplicity of the overall computation stems from the fact that in all steps of the algorithm only linear programs need to be solved.

9.1 Introduction

We have seen in Part II that for constrained *piecewise affine* (PWA) systems the *constrained finite time optimal control* (CFTOC) problem can be solved either by means of multi-parametric mixed-integer programming or (more efficiently) by utilizing a dynamic programming strategy combined with multi-parametric programming solvers. The resulting closed-form solution is a time-varying piecewise affine state feedback control law.

However, when a *receding horizon control* strategy is employed stability and feasibility for all time (Definition 3.4) of the closed-loop system are not guaranteed. To remedy this deficiency a priori various schemes have been proposed in the literature. For constrained linear systems stability can be (artificially) enforced by introducing the proper terminal set constraints and/or terminal cost function to the formulation of the CFTOC problem [MRRS00]. For the class of constrained PWA systems very few and restrictive stability criteria are known, e.g. [BM99a, MRRS00]. Only recently ideas, used for enforcing closed-loop stability of the CFTOC problem for constrained linear systems, have been extended to PWA systems [GKBM05]. Unfortunately the technique presented in [GKBM05] introduces a certain level of suboptimality in the solution.

Another way to guarantee closed-loop stability for PWA systems for the whole feasible state space is to attain a solution to the infinite time optimal control problem, as presented in Chapter 7.

In this chapter the focus is on the *a posteriori* analysis of the CFTOC solution, and the goal is to extract the regions of the state space for which closed-loop stability and feasibility for all time can be guaranteed. A technique to compute the maximal positively invariant set and a Lyapunov stability region (based on the linear cost function) for constrained PWA systems is presented. The algorithm combines a reachability analysis with some basic polyhedral manipulations. In the end of the chapter the applicability of the proposed algorithms is illustrated with several numerical examples.

9.2 Constrained Finite Time Optimal Control of Piecewise Affine Systems

Let us consider again the CFTOC Problem 6.1 from Chapter 6

$$J_T^*(x(0)) := \min_{U_T} \quad \|Px(T)\|_p + \sum_{t=0}^{T-1} \|Qx(t)\|_p + \|Ru(t)\|_p \quad \text{(6.1a)}$$

$$\text{subj. to } \begin{cases} x(t+1) = f_{\text{PWA}}(x(t), u(t)), \\ x(T) \in \mathcal{X}^f, \end{cases} \quad \text{(6.1b)}$$

with $p \in \{1, \infty\}$ and $U_T := \{u(t)\}_{t=0}^{T-1}$ for the piecewise affine (PWA) system

$$x(t+1) = f_{\text{PWA}}(x(t), u(t))$$
$$= A_d x(t) + B_d u(t) + a_d, \quad \text{if } \begin{bmatrix} x(t) \\ u(t) \end{bmatrix} \in \mathcal{D}_d, \quad \text{(4.1)}$$

where $\mathcal{D} := \cup_{d=1}^{N_\mathcal{D}} \mathcal{D}_d \subseteq \mathbb{R}^{n_x+n_u}$ is the domain of $f_{\text{PWA}}(\cdot, \cdot)$, which is non-empty, and $\{\mathcal{D}_d\}_{d=1}^{N_\mathcal{D}}$ denotes a polyhedral partition of the domain $\mathcal{D}$. The reader is referred to Chapter 4 for a detailed description of the piecewise affine system.

As shown in Chapter 6, the abovementioned problem can be solved using (for example) a dynamic programming strategy. The resulting optimal closed-form solution is of the following form: the value function is given by

$$J_k^*(x(t)) = \Phi_{k,i} x(t) + \Gamma_{k,i} \quad \text{if} \quad x(t) \in \mathcal{P}_{k,i}, \tag{6.11}$$

and the optimal input $u^*(t)$ is a time-varying piecewise affine function of the state $x(t)$, i.e. it is given as a *closed-form state feedback control law*

$$u^*(t) = \mu_k^*(x(t)) = F_{k,i} x(t) + G_{k,i}, \quad \text{if} \quad x(t) \in \mathcal{P}_{k,i}, \tag{6.12}$$

where $k = 1, \ldots, T$ with $t = T - k$, and $\{\mathcal{P}_{k,i}\}_{i=1}^{N_k}$ is a polyhedral partition of the set of feasible states $x(t)$ at time t

$$\mathcal{X}_k = \cup_{i=1}^{N_k} \mathcal{P}_{k,i},$$

with the closure of $\mathcal{P}_{k,i}$ given by $\bar{\mathcal{P}}_{k,i} = \{x \in \mathbb{R}^{n_x} \mid P_{k,i}^x x \leq P_{k,i}^0\}$.

When a *receding horizon* (RH) control (Chapter 3) policy is used, the control law turns out to be a time-invariant state feedback control law of the form

$$\mu_{\text{RH}}(x(t)) := \mu_T^*(x(t)) = F_{T,i} x(t) + G_{T,i}, \quad \text{if} \quad x(t) \in \mathcal{P}_{T,i}, \tag{6.14}$$

with $u(t) = \mu_{\text{RH}}(x(t))$ and a time-invariant cost function[1]

$$J_{\text{RH}}(x(t)) := J_T^*(x(t)) = \Phi_{T,i} x(t) + \Gamma_{T,i}, \quad \text{if} \quad x(t) \in \mathcal{P}_{T,i}, \tag{6.15}$$

for $t \geq 0$ and thus only $N_{\text{RH}} := N_T$ (in the worst case different) control law expressions have to be stored.

Remark 9.1 (Simplified notation). To simplify the notation in the rest of this chapter, let us discard the subscript T for the matrices and sets of region $i = 1, \ldots, N_{\text{RH}}$ in the receding horizon solution, i.e. $F_i := F_{T,i}, G_i := G_{T,i}, \mathcal{P}_i := \mathcal{P}_{T,i}, \Phi_i := \Phi_{T,i}, \Gamma_i := \Gamma_{T,i}, N := N_{\text{RH}} = N_T$, and $\mathcal{X} := \mathcal{X}_T$. □

9.3 Closed-Loop Stability and Feasibility

As mentioned in the beginning of this chapter, even for unconstrained linear systems, using receding horizon control without extending or modifying the underlying optimization problem, does *not* necessarily a priori guarantee closed-loop stability nor feasibility for all time (Definition 3.4) for the whole

---

[1] Note that $J_{\text{RH}}(\bullet)$ in (6.15) in general does not represent the value function of the *closed-loop* system when the receding horizon control law $\mu_{\text{RH}}(\bullet)$ is applied.

9.3 CLOSED-LOOP STABILITY AND FEASIBILITY

(initial open-loop) feasible state space $\mathcal{X}$, cf. [Löf03, MRRS00]. In his PhD thesis Johan Löfberg [Löf03, Example 2.1] for example shows that for a linear unconstrained system with a quadratic performance index the set in the (T, R)-parameter space, for which the closed-loop system is stable, is (possibly) a non-convex and disconnected set. Furthermore, the receding horizon control law might drive the system state outside of $\mathcal{X}$. Therefore, guaranteeing closed-loop stability and feasibility for all time for constrained PWA systems for the whole set $\mathcal{X}$ is far from being a simple task.

Assumption 9.2 (Unique dynamic). Without loss of generality one may assume that for all $i \in \{1, \ldots, N\}$, $\exists! \; d^\star = d(i)$ with $d^\star \in \{1, \ldots, N_\mathcal{D}\}$ such that for all $x \in \mathcal{P}_i$, $\left[\begin{smallmatrix} x \\ \mu_{\text{RH}}(x) \end{smallmatrix} \right] \in \mathcal{D}_{d^\star}$. □

Assumption 9.2 guarantees that the closed-loop system trajectories are uniquely defined. Note that this assumption is always fulfilled if the CFTOC solution is obtained with the dynamic programming based procedure described in Chapter 6. Otherwise it is possible to split the regions $\mathcal{P}_i$ further until Assumption 9.2 is met.

Having the receding horizon control law (6.14) at hand, the *autonomous closed-loop* (CL) system is given by

$$x(t+1) = f_{\text{PWA}}^{\text{CL}}(x(t)) := A_i^{\text{CL}} x(t) + a_i^{\text{CL}}, \quad \text{if } x(t) \in \mathcal{P}_i, \quad (9.1a)$$

where

$$A_i^{\text{CL}} := A_{d(i)} + B_{d(i)} F_i, \quad a_i^{\text{CL}} := a_{d(i)} + B_{d(i)} G_i, \quad i = 1, \ldots, N. \quad (9.1b)$$

Most of the computations in this chapter are based on the following essential concept of positively invariant sets, confer Definitions 2.2 and 2.3.

Definition 9.3 (Maximal positively invariant set $\mathcal{O}_{\max}$). Let $\mu_{\text{RH}}(\cdot)$, as in Equation (6.14), be a given control law for the PWA system (4.1). The largest set of states $\mathcal{O}_{\max} \subseteq \mathcal{X} \subseteq \mathbb{R}^{n_x}$ with

$$\mathcal{O}_{\max} := \left\{ x(0) \in \mathcal{X} \;\middle|\; x(t+1) = f_{\text{PWA}}^{\text{CL}}(x(t)) \in \mathcal{X}, \forall t \geq 0 \right\}$$

is called *maximal positively invariant set* for the closed-loop system (9.1). □

It follows immediately from the above definition that closed-loop *feasibility for all time*, i.e. the existence of a control strategy for all time, can be guaranteed if and only if one can confirm that the initial state belongs to the maximal positively invariant set $\mathcal{O}_{\max}$ [Bla99].

Note that it can be shown that for a given infinite horizon solution (if such a solution exists), i.e. the CITOC solution, the maximal positively invariant set is equal to the set of closed-loop (asymptotically) stable states which is in turn equal to the set $\mathcal{X}$, cf. Chapter 7. However, obtaining the CITOC solution for linear or PWA systems might be computationally prohibitive or demanding due to a large (possibly infinite) number of dynamic programming iterations and the complexity of the optimal solution itself. In the worst case the complexity (in the number of defining regions) of the problem might increase exponentially with increasing prediction horizon.

In many cases the numerical computation of the CITOC solution might not be possible or might not even be desired due to the complexity of the solution. As observed in [GLPM03], where a quadratic cost function for constrained linear systems is used, one can often neglect the difference in performance between the suboptimal CFTOC solution with a sufficiently large prediction horizon and the optimal CITOC solution but (most possibly) gains a tremendous complexity reduction. This behavior is very likely to be expected also for most (if not all) constrained PWA systems with a linear performance index, confer also Figure 9.11 and the example Section 9.4.2 itself. It is of major importance, however, to know for which subset of the open-loop feasible region $\mathcal{X}$ the computed suboptimal controller can guarantee closed-loop stability and feasibility.

9.3.1 Computation of the Maximal Positively Invariant Set

In order to present the algorithm, the following definitions are needed.

Definition 9.4 (Region of attraction $\mathcal{A}$). Let $\mu_{RH}(\cdot)$, as in Equation (6.14), be a given control law for the PWA system (4.1). The set of initial states $\mathcal{A} \subseteq \mathcal{X} \subseteq \mathbb{R}^{n_x}$, with

$$\mathcal{A} := \left\{ x(0) \in \mathcal{X} \mid \lim_{t \to \infty} x(t) = \mathbb{0}_{n_x} \right\}$$

is the *region of attraction* (to the origin) for the closed-loop system (9.1), cf. also Definition 2.6. □

From the Definitions 9.3 and 9.4 it immediately follows that $\mathcal{A} \subseteq \mathcal{O}_{max}$. However, the complementary set $\mathcal{O}_{max} \setminus \mathcal{A}$ for the class of autonomous piecewise affine systems (9.1) can basically incorporate any system behavior, such as limit cycles, stationary points (other than the origin), general attractive manifolds and trajectories that lead to these attractors, or even chaotic behavior as illustrated by the simple tent map [Dav03, Wik].

Definition 9.5 (Lyapunov stability region $\mathcal{V}$). Let $V : \mathcal{X} \to \mathbb{R}_{\geq 0}$ be some real-valued function and $\mu_{RH}(\cdot)$, as in Equation (6.14), be a given control law for the PWA system (4.1). The set of initial states $\mathcal{V} \subseteq \mathcal{X} \subseteq \mathbb{R}^{n_x}$, with

$$\mathcal{V} := \left\{ x(0) \in \mathcal{X} \mid V(f_{PWA}^{CL}(x(t))) < V(x(t)),\ x(t) \in \mathcal{X},\ \forall t \geq 0 \right\}$$

is called (*set* or) *region of Lyapunov stability* for the closed-loop system (9.1) □

9.3 CLOSED-LOOP STABILITY AND FEASIBILITY

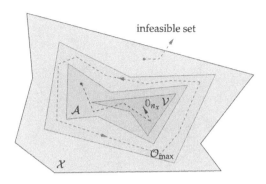

Fig. 9.1: Possible arrangement of the open-loop feasible set $\mathcal{X}$, the maximal positively invariant set $\mathcal{O}_{\max}$, the region of attraction $\mathcal{A}$, and some Lyapunov stability region $\mathcal{V}$. The dashed lines depict possible trajectories starting in the respective sets.

Note that $\mathcal{V}$ is a conservative indicator for the set of initial states with closed-loop asymptotic stability based on a Lyapunov function candidate $V(\cdot)$.

Figure 9.1 shows a typical arrangement of the open-loop feasible set $\mathcal{X}$, the maximal positively invariant set $\mathcal{O}_{\max}$, the region of attraction $\mathcal{A}$, some Lyapunov stability region $\mathcal{V}$ as well as the typical behavior for a trajectory $x(t)$ starting in these respective sets (dashed lines) for constrained PWA systems.

In [GLPM03] the authors compute the invariant set in an iterative procedure where at each iteration step a one-step reachability analysis is performed to extract the parts of the state space that remain closed-loop feasible. The algorithm has converged when the feasible state space remains constant. The approach presented here to compute the maximal positively invariant set $\mathcal{O}_{\max}$ for a given PWA state feedback control law for constrained PWA systems can be considered as complementary to the algorithm presented in [GLPM03].

The here considered approach also uses an iterative approach but, in contrast to [GLPM03], the focus here is on the computation of the parts of the open-loop feasible state space that lead to 'infeasibility regions', denoted with $\mathcal{U}_i$. One of the benefits of this approach is that there is no need for the computation of the union of polyhedral regions in intermediate steps. The detailed procedure of the maximal positively invariant set computation is given in Algorithm 9.6. Potential speed-ups of the algorithm are not mentioned here. Note that for simplicity in the following only operations on (possibly non-convex) sets and polyhedral regions are mentioned but in fact to every 'region' a cost function (6.15) and state feedback control law (6.14), respectively, are assigned.

The algorithm is divided into an initialization part and a main part. A schematical arrangement of the considered regions and sets in iteration step $[r]$ used in Algorithm 9.6 is depicted in Figure 9.2. The dashed arrow denotes the

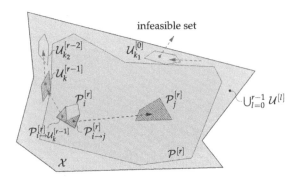

Fig. 9.2: Schematical arrangement of the regions being used in Algorithm 9.6 in iteration step $[r]$. The dashed arrow denotes that the target set is reached in one step with the given control law.

reachability from the 'source'-set to the 'target'-set in one step using the given PWA state feedback control law.

In the *initialization part* a one-step reachability analysis is performed and the possible mappings from the polyhedral region $\mathcal{P}_i$ to the region $\mathcal{P}_j$ are recorded in the mapping matrix M, i.e. if parts of the region $\mathcal{P}_j$ can be reached from parts of region $\mathcal{P}_i$ in one step by the given piecewise affine control law (6.14) then the entry $M_{i \to j}$ in the mapping matrix is set to **true**. See Figure 9.2 for a schematical explanation. $\mathcal{P}_{i \to j}$ denotes the part of $\mathcal{P}_i$ that is mapped into region $\mathcal{P}_j$ in one step. Additionally, regions that lead to infeasibility, denoted with $\mathcal{U}_k^{[0]}$, in one step are computed.

In the *main part* at every iteration step $[r]$, a one-step reachability analysis from the 'feasible' region $\mathcal{P}_i^{[r]}$, where $i = 1, \ldots, N_\mathcal{P}^{[r]}$, to the *'infeasible' regions*[2] $\mathcal{U}_k^{[r-1]}$, where $k = 1, \ldots, N_\mathcal{U}^{[r-1]}$, from the previous iteration step is performed. The part of $\mathcal{P}_i^{[r]}$ that is mapped into the infeasible region $\mathcal{U}_k^{[r-1]}$ in one step is denoted with $\mathcal{P}_{i \to \mathcal{U}_k^{[r-1]}}^{[r]}$, cf. Figure 9.2. The union of all $\mathcal{P}_{i \to \mathcal{U}_k^{[r-1]}}^{[r]}$ is the new 'infeasibility' set $\mathcal{U}^{[r]}$. At the end of every iteration step $[r]$, the set difference of the feasible set $\mathcal{P}^{[r-1]}$ from the beginning of the iteration step $[r]$ and the newly computed closed-loop infeasible set $\mathcal{U}^{[r]}$ of the state space is performed. This set-difference is the 'new' feasible set $\mathcal{P}^{[r]}$.

The algorithm has converged when no regions of the feasible set are leading to infeasibility, i.e. when for some iteration step r^* we have $\mathcal{U}^{[r^*]} = \emptyset$. The remaining feasible set is the maximal positively invariant set $\mathcal{O}_{\max}$ because all states starting in this remaining set will remain in $\mathcal{O}_{\max}$ for all time by construction.

---

[2] 'Infeasible' regions $\mathcal{U}^{[r-1]}$ denote all the states in $\mathcal{X}$ that are driven into the infeasibility set in r steps.

9.3 Closed-Loop Stability and Feasibility

Note that, by design, every region $\mathcal{P}_j^{[r]}$ is a subset of some original region $\mathcal{P}_i$. Therefore, the control law and cost expressions of $\mathcal{P}_j^{[r]}$ are the same as for the original region but the region index of the subsets is changed from iteration step to iteration step. (The same applies for $\mathcal{U}_k^{[r]}$.) Thus a simple but very efficient speed-up is performed by using the matrix of the possible transitions (mappings) M, computed in the initialization step, and the 'region index' function $\text{org}(\cdot) : \mathcal{X} \to \mathbb{N}$, defined implicitly as

$$\text{org}(\mathcal{P}_j^{[r]}) = i \quad \Leftrightarrow \quad \mathcal{P}_j^{[r]} \subseteq \mathcal{P}_i,$$

thus $\text{org}(\mathcal{P}_j^{[r]}) = i$ denotes the index i of the original polyhedral region $\mathcal{P}_i$ of the original partitioning $\mathcal{P}$, where $\mathcal{P}_j^{[r]}$ originates from in iteration step r. Therefore, $\mathcal{P}_j^{[r]} \subseteq \mathcal{P}_i$.

Thus it is easy to check in iteration step $[r]$ which regions give a possibility to lead into feasibility or infeasibility in one step, and which transitions are impossible with the computed control law, e.g. if

$$M_{\text{org}(\mathcal{P}_j^{[r]}) \to \text{org}(\mathcal{U}_k^{[r-1]})} = \textsf{false}$$

then the subset of $\mathcal{P}_j^{[r]}$ that will lead the state in one step into $\mathcal{U}_k^{[r-1]}$ is empty because it was computed in the initialization step that there is no one step transition from the original region $\mathcal{P}_i$ to $\mathcal{P}_{\text{org}(\mathcal{U}_k^{[r-1]})}$. This reduces the number of possible combinations tremendously as the iterations evolve.

From the description above it is clear that an efficient computation of the set difference has a major impact on the implementation of the algorithm. In this work the set difference is computed with the procedure presented in [Bao05, Sec. 3.3], since it involves only linear programs and the number of regions it generates for the description of the set difference is very low. (See [Bao05, Sec. 3.3] for more details.)

Algorithm 9.6 (Maximal positively invariant set $\mathcal{O}_{\max}$) _____

INPUT   CFTOC solution: $f_{\text{PWA}}(x, u)$, $\mu_{\text{RH}}(x)$, $J_{\text{RH}}(x)$, $\{\mathcal{P}_i\}_{i=1}^N$
OUTPUT Maximal positively invariant set $\mathcal{O}_{\max}$

_____ Initialization _____

$\mathcal{U}^{[0]} := \emptyset$
FOR $i = 1$ **TO** N
    compute A_i^{CL}, a_i^{CL} according to (9.1b)
    FOR $j = 1$ **TO** N

$$\mathcal{P}_{i \to j} := \left\{ x \in \mathbb{R}^{n_x} \middle| \begin{bmatrix} P_i^x \\ P_j^x A_i^{\text{CL}} \end{bmatrix} x \leq \begin{bmatrix} p_i^0 \\ p_j^0 - P_j^x a_i^{\text{CL}} \end{bmatrix} \right\}$$

    IF $\mathcal{P}_{i \to j} \neq \emptyset$
        THEN $M_{i \to j} :=$ **true**, **ELSE** $M_{i \to j} :=$ **false**, **END**

    IF $\mathcal{P}_i \backslash \bigcup_{k=1}^{j} \mathcal{P}_{i \to k} = \emptyset$
        THEN $M_{i \to k} :=$ **false** for $k = j+1, \ldots, N$; **NEXT** i, **END**
END
$$\mathcal{U}^{[0]} := \mathcal{U}^{[0]} \cup \left(\mathcal{P}_i \backslash \bigcup_{k=1}^{N} \{\mathcal{P}_{i \to k}\} \right)$$
END

_____ MAIN _____

$r := 0, \quad \mathcal{U}^{[0]} =: \bigcup_{j=1}^{N_u^{[r]}} \{u_j^{[r]}\}, \quad \mathcal{P}^{[r]} := \mathcal{P} \backslash \mathcal{U}^{[0]} = \bigcup_{i=1}^{N_\mathcal{P}^{[r]}} \{\mathcal{P}_i^{[r]}\}$

WHILE $\mathcal{U}^{[r]} \neq \emptyset$

  $r := r + 1, \quad \mathcal{U}^{[r]} := \emptyset$

  FOR $i = 1$ **TO** $N_\mathcal{P}^{[r-1]}$

    $\mathcal{M} := \left\{ k \; \middle| \; M_{\mathrm{org}(\mathcal{P}_i^{[r-1]}) \to \mathrm{org}(\mathcal{U}_k^{[r-1]})} = \text{true} \right\}$

    FOR EACH $j \in \mathcal{M}$

$$\mathcal{P}_{i \to u_j^{[r-1]}}^{[r-1]} := \left\{ x \in \mathbb{R}^{n_x} \; \middle| \; \begin{bmatrix} P_i^{x,[r-1]} \\ U_j^{x,[r-1]} A_{\mathrm{org}(\mathcal{P}_i^{[r-1]})}^{\mathrm{CL}} \end{bmatrix} x \leq \begin{bmatrix} p_i^{0,[r-1]} \\ u_j^{0,[r-1]} - U_j^{x,[r-1]} a_{\mathrm{org}(\mathcal{P}_i^{[r-1]})}^{\mathrm{CL}} \end{bmatrix} \right\},$$

      $\mathcal{U}^{[r]} := \mathcal{U}^{[r]} \cup \mathcal{P}_{i \to u_j^{[r-1]}}^{[r-1]}$

      IF $\mathcal{P}_i^{[r-1]} = \bigcup_{k=\mathcal{M}_1}^{j} \left\{ \mathcal{P}_{i \to u_k^{[r-1]}}^{[r-1]} \right\}$ **THEN NEXT** i, **END**

    END

  END

$\mathcal{U}^{[r]} =: \bigcup_{j=1}^{N_u^{[r]}} \{u_j^{[r]}\},$

$\mathcal{P}^{[r]} := \bigcup_{i=1}^{N_\mathcal{P}^{[r-1]}} \{\mathcal{P}_i^{[r-1]}\} \backslash \mathcal{U}^{[r]} = \bigcup_{i=1}^{N_\mathcal{P}^{[r]}} \{\mathcal{P}_i^{[r]}\}$

END

$\mathcal{O}_{\max} := \mathcal{P}^{[r]}$

Remark 9.7 (Finite termination). Note that there is no guarantee for finite termination of the Algorithm 9.6 for computing the maximal positively invariant set $\mathcal{O}_{\max}$, even for constrained linear systems. And in the case that the open-loop feasible set $\mathcal{X}$ is open, the Algorithm for computing $\mathcal{O}_{\max}$ takes an infinite number of iteration steps. However, this is hardly a limitation for 'practical' problems, where usually a δ-close compact approximation $\mathcal{X}^\delta \subset \mathcal{X}$ is considered. □

9.3.2 Computation of the Lyapunov Stability Region

Next an algorithm to compute a Lyapunov stability region $\mathcal{V}$ as in Definition 9.5 for a given CFTOC solution, when the receding horizon control strategy is applied, is presented. The algorithm is based on the linear cost function $J_{RH}(\cdot)$ and additionally uses reachability analysis for the computation. Unlike techniques presented in the literature, cf. e.g. [FTCMM02, JR98], we are not looking for a piecewise quadratic Lyapunov function that provides overall stability guarantees but are checking if the given value function of the CFTOC solution is a piecewise linear Lyapunov function for (at least) a subset of the feasible state space. Here no LMI techniques are needed, and no possible additional conservatism is introduced.

In analogy to 'energy' arguments it is natural to assume that the PWA value function (6.15) is a valid candidate for a PWA Lyapunov function for a region around the origin, cf. also Chapter 7. In contrast to the optimal infinite time solution where the whole feasible state space is stabilizing, the finite time solution with a receding horizon control strategy is considered. Therefore, it is not to be expected (especially for PWA systems) that with standard Lyapunov stability arguments with this candidate of a Lyapunov function, that the whole asymptotically stable region of the closed-loop system is obtained, but only a subset of it. In the unfortunate case, when the value function (6.15) behaves like a Lyapunov function only on non-full-dimensional parts of the state space, such as e.g. solely the origin, the proposed algorithm will fail. (See Remark 9.9 for a possible practical remedy.) However, even though not explicitly stated here, the extension to the non-full-dimensional case applies similarly.

The detailed description is given in Algorithm 9.8. The notation is analog to the notation used in Algorithm 9.6 for computing the maximal positively invariant set $\mathcal{O}_{max}$. Also here, note that for simplicity in the following only operations on (possibly non-convex) sets and polyhedral regions are mentioned; however, it is also needed to keep track of the cost function and state feedback control law (as in (6.15) and (6.14), respectively) assigned to each region. Speed-ups of the algorithm are again not discussed for simplicity.

The *initialization part* of the algorithm is completely analog to the initialization part of Algorithm 9.6 but in addition the 'Lyapunov' stability set $\mathcal{V}^{[0]}$ is computed where in one step the cost function $J_{RH}(\cdot)$, which is the Lyapunov candidate, is decreased, i.e.

$$\mathcal{V}^{[0]} := \{x \in \mathcal{X} \mid J_{RH}(f_{PWA}^{CL}(x)) < J_{RH}(x), f_{PWA}^{CL}(x) \in \mathcal{X}\}.$$

Starting with $\mathcal{V}^{[0]}$ as initialization in the *second part* of the algorithm, the parts of $\mathcal{V}^{[0]}$ are extracted for which the value function decays in one step. The remaining part is the new set $\mathcal{V}^{[1]}$ and the procedure continues in an iterative way until in two consecutive steps the polyhedral partition of the Lyapunov

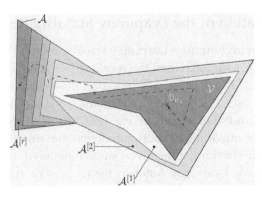

Fig. 9.3: Possible arrangement of some Lyapunov stable region $\mathcal{V}$, the reach-sets $\mathcal{A}^{[r]}$, and the region of attraction to the origin $\mathcal{A}$. The dashed lines depicts a possible trajectory starting in the region of attraction.

stability region does not change, i.e. $\mathcal{V}^{[r]} = \mathcal{V}^{[r+1]} =: \mathcal{V}$. Note that by construction the remaining set $\mathcal{V}$ is a Lyapunov stable and positively invariant set but not necessarily maximal positively invariant.

Third part: To provide a good 'fix' for the aforementioned deficiency of the Lyapunov function of our choice, i.e. that the whole asymptotic stability region is (possibly) not covered, a one-step reachability analysis into the Lyapunov region $\mathcal{V}$ is performed. The newly attained regions are denoted with $\mathcal{A}_i^{[1]}$. Confer also Figure 9.3. Then, in an iterative way, again a one-step reachability analysis into the regions $\mathcal{A}_k^{[r-1]}$ is performed to obtain the regions $\mathcal{A}_i^{[r]}$ until no new regions are be found. It is clear from the construction that the set $\mathcal{A}^{[r]}$ is the set of states from which a trajectory is driven (with the given control law) into the computed Lyapunov stable region $\mathcal{V}$ in r steps, i.e. $\mathcal{A}^{[r]}$ is a region of attraction to $\mathcal{V}$ (Definition 2.6). The union of all $\mathcal{A}^{[r]}$ together with the Lyapunov stability set $\mathcal{V}$ is the region of attraction $\mathcal{A}$ to the origin, which itself is a Lyapunov stability set in the classical sense.

Algorithm 9.8 (Lyapunov stability region $\mathcal{V}$, region of attraction $\mathcal{A}$)

INPUT CFTOC solution: $f_{\text{PWA}}(x, u)$, $\mu_{\text{RH}}(x)$, $J_{\text{RH}}(x)$, $\{\mathcal{P}_i\}_{i=1}^N$

OUTPUT Lyapunov stability region $\mathcal{V}$, region of attraction $\mathcal{A}$

―――――――――― INITIALIZATION ――――――――――

$\mathcal{V}^{[0]} := \emptyset$

FOR $i = 1$ **TO** N
  compute A_i^{CL}, a_i^{CL}, according to (9.1b)
  FOR $j = 1$ **TO** N^0

$$\mathcal{P}_{i \to j} := \left\{ x \in \mathbb{R}^{n_x} \;\middle|\; \begin{bmatrix} p_i^x \\ p_j^x A_i^{\text{CL}} \end{bmatrix} x \leq \begin{bmatrix} p_i^0 \\ p_j^0 - p_j^x a_i^{\text{CL}} \end{bmatrix} \right\}$$

9.3 Closed-Loop Stability and Feasibility

$\quad$ IF $\quad \mathcal{P}_{i \to j} \neq \emptyset$ THEN
$\quad\quad M_{i \to j} :=$ true,
$\quad\quad \mathcal{S}^{[0]}_{i \to j} := \left\{ x \in \mathbb{R}^{n_x} \mid \left(\Phi_j A^{\text{CL}}_i - \Phi_i \right) x < \left(\Gamma_i - \Gamma_j - \Phi_j A^{\text{CL}}_i \right) \right\}$,
$\quad\quad \mathcal{V}^{[0]} := \mathcal{V}^{[0]} \cup \left(\mathcal{P}_{i \to j} \cap \mathcal{S}^{[0]}_{i \to j} \right)$
$\quad$ ELSE $\;\; M_{i \to j} =$ false, END
$\quad$ IF $\quad \mathcal{P}_i = \cup_{k=1}^{j} \mathcal{P}_{i \to k}$
$\quad\quad$ THEN $\; M_{i \to k} =$ false $\;$ for $k = j+1, \dots, N$; NEXT i, END
END
END

$\quad\quad\quad\quad\quad\quad\quad\quad$ _____ Lyapunov Stability Region $\mathcal{V}$ _____

$r := 0, \;\; \mathcal{V}^{[0]} =: \bigcup_{j=1}^{N^{[r]}_\mathcal{V}} \left\{ \mathcal{V}^{[r]}_j \right\}, \;\; \Delta \mathcal{V}^{[0]} := \mathcal{P} \backslash \mathcal{V}^{[0]}$

WHILE $\Delta \mathcal{V}^{[r]} \neq \emptyset$
$\quad r := r+1, \;\; \mathcal{V}^{[r]} := \emptyset, \;\; \Delta \mathcal{V}^{[r]} := \emptyset$
$\quad$ **FOR** $i = 1$ **TO** $N^{[r-1]}_\mathcal{V}$
$\quad\quad \mathcal{M} := \left\{ k \;\middle|\; M_{\text{org}(\mathcal{V}^{[r-1]}_i) \to \text{org}(\mathcal{V}^{[r-1]}_k)} =$ true $\right\}$

$\quad\quad$ **FOR EACH** $j \in \mathcal{M}$
$\quad\quad\quad \mathcal{V}^{[r]}_{i \to j} := \left\{ x \in \mathbb{R}^{n_x} \;\middle|\; \begin{bmatrix} V^{x,[r-1]}_i \\ V^{x,[r-1]}_j A^{\text{CL}}_{\text{org}(\mathcal{V}^{[r-1]}_i)} \end{bmatrix} x \leq \begin{bmatrix} V^{0,[r-1]}_i \\ V^{0,[r-1]}_j - V^{x,[r-1]}_j a^{\text{CL}}_{\text{org}(\mathcal{V}^{[r-1]}_i)} \end{bmatrix} \right\}$,

$\quad\quad\quad$ IF $\;\; \mathcal{V}^{[r]}_{i \to j} \neq \emptyset$ THEN
$\quad\quad\quad\quad \mathcal{S}^{[r]}_{i \to j} := \left\{ x \in \mathbb{R}^{n_x} \;\middle|\; \left(\Phi_j A^{\text{CL}}_{\text{org}(\mathcal{V}^{[r-1]}_i)} - \Phi_i \right) x \right.$
$\quad\quad\quad\quad\quad\quad\quad\quad\quad\quad\quad\quad\quad\quad\quad\quad\quad\quad\quad\quad \left. < \left(\Gamma_i - \Gamma_j - \Phi_j A^{\text{CL}}_{\text{org}(\mathcal{V}^{[r-1]}_i)} \right) \right\}$,
$\quad\quad\quad\quad \mathcal{V}^{[r]} := \mathcal{V}^{[r]} \cup \left(\mathcal{V}^{[r]}_{i \to j} \cap \mathcal{S}^{[r]}_{i \to j} \right)$
$\quad\quad\quad$ END
$\quad\quad\quad$ IF $\;\; \mathcal{V}^{[r-1]}_i = \cup_{k=\mathcal{M}_1}^{j} \left\{ \mathcal{V}^{[r]}_{i \to k} \right\}$ THEN $\;$ NEXT i, END
$\quad\quad$ END
$\quad$ END
$\mathcal{V}^{[r]} \;\; =: \bigcup_{j=1}^{N^{[r]}_\mathcal{V}} \left\{ \mathcal{V}^{[r]}_j \right\}$

$\Delta \mathcal{V}^{[r]} := \mathcal{V}^{[r-1]} \setminus \mathcal{V}^{[r]}$
END
$\mathcal{V} := \mathcal{V}^{[r]}$

────────── REGION OF ATTRACTION $\mathcal{A}$ ──────────

$r := 0, \quad \mathcal{A} := \mathcal{V}, \quad \mathcal{A}^{[0]} := \mathcal{V} = \bigcup_{j=1}^{N_\mathcal{A}^{[0]}} \{A_j^{[0]}\}, \quad \mathcal{R}^{[0]} := \mathcal{P} \setminus \mathcal{V} = \bigcup_{j=1}^{N_\mathcal{R}^{[0]}} \{R_j^{[0]}\}$

WHILE $\mathcal{A}^{[r]} \neq \emptyset$ **AND** $\mathcal{R}^{[r]} \neq \emptyset$

$\quad r := r+1, \quad \mathcal{A}^{[r]} := \emptyset$

$\quad$ **FOR** $i = 1$ **TO** $N_\mathcal{R}^{[r-1]}$

$\quad\quad \mathcal{M} := \left\{ k \;\middle|\; M_{\mathrm{org}(R_i^{[r-1]}) \to \mathrm{org}(A_k^{[r-1]})} = \text{true} \right\}$

$\quad\quad$ **FOR EACH** $j \in \mathcal{M}$

$\quad\quad\quad \mathcal{R}_{i \to A_j^{[r-1]}}^{[r-1]} := \left\{ x \in \mathbb{R}^{n_x} \;\middle|\; \begin{bmatrix} R_i^{x,[r-1]} \\ A_j^{x,[r-1]} A_{\mathrm{org}(R_i^{[r-1]})}^{\mathrm{CL}} \end{bmatrix} x \leq \begin{bmatrix} R_i^{0,[r-1]} \\ A_j^{0,[r-1]} - A_j^{x,[r-1]} A_{\mathrm{org}(R_i^{[r-1]})}^{\mathrm{CL}} \end{bmatrix} \right\}$,

$\quad\quad\quad \mathcal{A}^{[r]} := \mathcal{A}^{[r]} \cup \mathcal{R}_{i \to A_j^{[r-1]}}^{[r-1]}$

$\quad\quad\quad$ **IF** $\mathcal{R}_i^{[r-1]} = \bigcup_{k=\mathcal{M}_1}^{j} \left\{ \mathcal{R}_{i \to A_k^{[r-1]}}^{[r-1]} \right\}$ **THEN NEXT** i, **END**

$\quad\quad$ **END**

$\quad$ **END**

$\quad \mathcal{R}^{[r]} := \bigcup_{j=1}^{N_\mathcal{R}^{[r-1]}} \{R_j^{[r-1]}\} \setminus \mathcal{A}^{[r]} = \bigcup_{j=1}^{N_\mathcal{R}^{[r]}} \{R_j^{[r]}\}$,

$\quad \mathcal{A}^{[r]} =: \bigcup_{j=1}^{N_\mathcal{A}^{[r]}} \{A_j^{[r]}\}$

END
$\mathcal{A} := \bigcup_{k=0}^{r} \mathcal{A}^{[k]}$

Note that the computation of the union of regions in the various steps of Algorithm 9.6 and Algorithm 9.8 is very simple. Since the regions do not intersect, the union can simply be represented as a P-collection of these regions.

Remark 9.9 (Region of attraction). For the case that $0_{n_x} \in \mathrm{int}(\mathcal{V})$, the set $\mathcal{A}$ computed with Algorithm 9.8 is the (maximal) region of attraction as in

9.3 CLOSED-LOOP STABILITY AND FEASIBILITY

Definition 9.4. However, if $0_{n_x} \notin \text{int}(\mathcal{V})$ then in general Algorithm 9.8 computes a subset of the (maximal) region of attraction. See the example in Section 9.4.3 and the related Figure 9.10. In such a case one could either increase the prediction horizon T and check if for the new CFTOC solution $0_{n_x} \in \text{int}(\mathcal{V})$, or compute the region of attraction to an ε-size hypercube around the origin. □

Remark 9.10 (Open feasible set $\mathcal{X}$). As in Algorithm 9.6, it should be noted that in the case that the open-loop feasible set $\mathcal{X}$ is open, the Algorithm 9.8 will possibly not terminate in a finite number of iteration steps. However, under stronger assumptions (e.g. exponential stability) as it was shown in [GLPM03], a finite termination of such an algorithm can be guaranteed. For the considered case where no assumptions on the stability of the closed-loop trajectories are made, it is not to be expected that finite termination of the algorithm can be guaranteed. □

Theorem 9.11 (Asymptotic stability of $\mathcal{A}$). The closed-loop controlled system (9.1) is *asymptotically stabilizing* to the origin for every state $x(t)$ starting in (or traveling trough) the region of attraction $\mathcal{A}$. □

Proof. The proof follows by the construction of the region of attraction $\mathcal{A}$ (Algorithm 9.8) and standard Lyapunov stability arguments. As mentioned before, the set $\mathcal{A}$ is itself a Lyapunov stability region in the classical sense. ■

Corollary 9.12. Let $\mathcal{A}$ be the region of attraction and $\mathcal{O}_{\max}$ be the maximal positively invariant set of the closed-loop system (9.1). Then the following holds:

(a) If $\mathcal{A} \equiv \mathcal{O}_{\max}$ then all states $x(t) \in \mathcal{O}_{\max}$ will asymptotically converge to the origin, i.e. no limit cycles or stationary point/attractors other than the origin exist.

(b) If $\mathcal{A} \subset \mathcal{O}_{\max} \subseteq \mathbb{R}^{n_x}$ then stationary points/attractors other than the origin and/or limit cycles *can* lie in the set $\mathcal{O}_{\max} \setminus \mathcal{A}$. □

Proof. The proof of Corollary 9.12 follows directly from Theorem (9.11). ■

Theorem 9.13 (Infeasibility). Let $\mathcal{O}_{\max}$ be the maximal positively invariant set of the closed-loop system (9.1) and $\mathcal{X} \supseteq \mathcal{O}_{\max}$ with $\mathcal{X} \subseteq \mathbb{R}^{n_x}$ be the feasible set of the CFTOC Problem 6.1. Any state $x(t) \in (\mathcal{X} \setminus \mathcal{O}_{\max})$ with $t \geq 0$ will leave $\mathcal{X}$ for some t^*, i.e. there exits a $t^* \in \mathbb{Z}_{>0}$ such that $x(t) \notin \mathcal{X}$ for $t \geq t^*$ and $x(t) \in (\mathcal{X} \setminus \mathcal{O}_{\max})$ for $t^* > t \geq 0$. □

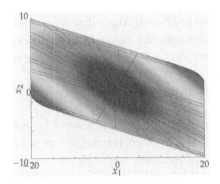
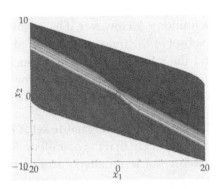

Fig. 9.4: Cost function $J_{RH}(x)$ for example (9.2) with $T = 5$. Same color implies same cost value.

Fig. 9.5: Control law $\mu_{RH}(x)$ for example (9.2) with $T = 5$. Blue ($u = -1$) and red-brown ($u = 1$) marked regions correspond to the 'saturated' control action.

Proof. The proof follows by the construction of the maximal positively invariant set a $\mathcal{O}_{max}$ and region of 'infeasibility' $\mathcal{U}$ (Algorithm 9.6). ∎

9.4 Examples

9.4.1 Constrained Double Integrator

Consider the constrained double integrator

$$\begin{cases} x(t+1) = \begin{bmatrix} 1 & 1 \\ 0 & 1 \end{bmatrix} x(t) + \begin{bmatrix} 1 \\ 1/2 \end{bmatrix} u(t), \\ x(t) \in [-20, 20] \times \mathbb{R}, \quad \text{and} \quad u(t) \in [-1, 1]. \end{cases} \quad (9.2)$$

The constrained finite time optimal control problem (6.1a)–(6.1b) is solved with $Q = I_2$, $R = 1$, $P = 0_{2\times 2}$, and $\mathcal{X}^f = [-20, 20] \times [-20, 20]$ for $p = \infty$. $T = 5$ was chosen as prediction horizon. The solution to the CFTOC problem is reported in Figure 9.4 (cost function $J_{RH}(\cdot)$), Figure 9.5 (receding horizon control law $\mu_{RH}(\cdot)$), and Table 9.1.

computation for	$T = 5$	$T = 10$
Lyapunov stability region $\mathcal{V}$	8 iterations	6 iterations
	178 regions	556 regions
region of attraction $\mathcal{A}$	11 iterations	8 iterations
	1 256 regions	928 regions
max. positively invariant set $\mathcal{O}_{max}$	7 iteration	4 iteration
	266 regions	444 regions
CFTOC solution	82 regions	446 regions

Tab. 9.1: Computational results for example (9.2).

9.4 EXAMPLES

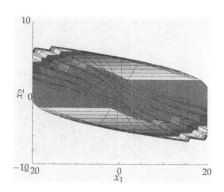

Fig. 9.6: Lyapunov stability region $\mathcal{V}$ (red) and region of attraction $\mathcal{A}$ (red+grey-shades) for example (9.2) with $T = 5$.

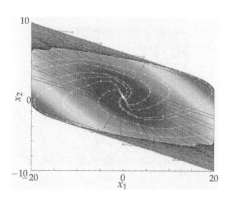

Fig. 9.7: Maximal positively invariant set $\mathcal{O}_{max}$ for example (9.2) with $T = 5$. Same color implies same cost value. Green marked regions lead to infeasibility and do not belong to the set $\mathcal{O}_{max}$.

For $T = 5$ the region of attraction $\mathcal{A}$ and the maximal positively invariant set $\mathcal{O}_{max}$ are identical, cf. Figure 9.6 and Figure 9.7. This means that no limit cycle or stationary points other than the origin exist. Furthermore, the stability region is a strict subset of the open-loop feasible region $\mathcal{X}$, i.e. there exist regions of the CFTOC solution for $T = 5$ which lead to infeasibility when the receding horizon policy is applied (green marked regions and red marked trajectories in Figure 9.7).

9.4.2 Constrained PWA Sine-Cosine System

Consider again the piecewise affine system [BM99a] from Chapter 6 and 7

$$\begin{cases} x(t+1) &= \frac{4}{5} \begin{bmatrix} \cos \alpha(x(t)) & -\sin \alpha(x(t)) \\ \sin \alpha(x(t)) & \cos \alpha(x(t)) \end{bmatrix} x(t) + \begin{bmatrix} 0 \\ 1 \end{bmatrix} u(t), \\ \alpha(x(t)) &= \begin{cases} \frac{\pi}{3} & \text{if } [1\ 0]x(t) \geq 0, \\ -\frac{\pi}{3} & \text{if } [1\ 0]x(t) < 0, \end{cases} \\ x(t) &\in [-10, 10] \times [-10, 10], \quad \text{and} \\ u(t) &\in [-1, 1]. \end{cases} \quad (6.16)$$

The constrained finite time optimal control problem (6.1a)–(6.1b) is solved with $Q = I_2$, $R = 1$, $P = 0_{2 \times 2}$, and $\mathcal{X}^f = [-10, 10] \times [-10, 10]$ for $p = \infty$. $T = 1$ was chosen as prediction horizon. The results are depicted in Figure 9.8–9.10 and in Table 9.2.

From Chapter 6 and 7 we know that the CITOC solution (comprising 252 polyhedral regions) for the system (6.16) is obtained with a prediction horizon $T = T_\infty = 11$. Nevertheless, as can be seen from Table 9.2 even for the

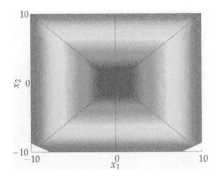

Fig. 9.8: Maximal positively invariant set $\mathcal{O}_{\max}$ for example (6.16) with $T = 1$. Same color corresponds to the same cost value $J_{RH}(x)$.

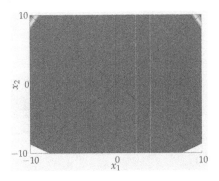

Fig. 9.9: Control law $\mu_{RH}(x)$ for example (6.16) with $T = 1$. Same color corresponds to the same control law expression.

small prediction horizon of $T = 1$ the maximal positively invariant set is $\mathcal{O}_{\max} = \mathcal{X} \subset \mathbb{R}^{n_x}$ which means that the overall system is stable. The set $\mathcal{X}$ for $T = 1$ is partitioned into only 10 polyhedral regions compared to the more complex CITOC solution with 252 polyhedral regions, cf. Figure 9.8. In Figure 9.10 it can be seen that the Lyapunov stability region $\mathcal{V}$, as computed in Section 9.3.2, does not cover $\mathcal{X}$ and therefore (in this case) the cost function is not the best candidate for a Lyapunov function. Note that $\mathbb{O}_{n_x} \notin \text{int}(\mathcal{V})$ which, as pointed out in Remark 9.9, means that Algorithm 9.8 may (and in this case does) return a subset of the (maximal) region of attraction as seen from the trajectory in Figure 9.10. By increasing the prediction horizon to $T = 2$ one obtains $\mathcal{X} = \mathcal{A}$, confer Remark 9.9.

The performance decay index $e_J(\cdot)$ and the maximum performance deviation $e_J^{\max}(\cdot)$ from the optimal infinite time solution as a function of the prediction horizon T are depicted in Figure 9.11. Both are computed as follows: the feasible state space is gridded, and for each equidistant grid point $x_{0,i}$ with

computation for	$T = 1$	$T = 3$
Lyapunov stability region $\mathcal{V}$	5 iterations 18 regions $\mathcal{V} \subset \mathcal{X}$	1 iteration 26 regions $\mathcal{V} = \mathcal{X}$
region of attraction $\mathcal{A}$	2 iterations 20 regions $\mathcal{A} \subset \mathcal{X}$	(1 iteration) (26 regions) $\mathcal{A} = \mathcal{V} = \mathcal{X}$
max. positively invariant set $\mathcal{O}_{\max}$	1 iteration 10 regions $\mathcal{O}_{\max} = \mathcal{X}$	1 iteration 26 regions $\mathcal{O}_{\max} = \mathcal{X}$
CFTOC solution	10 regions	26 regions

Tab. 9.2: Computational results for system (6.16).

9.4 Examples

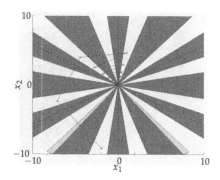
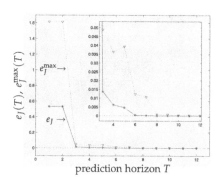

Fig. 9.10: Lyapunov stability region $\mathcal{V}$ (red) and a subset of the (maximal) region of attraction $\mathcal{A}$ (red+grey) for example (6.16) with $T = 1$.

Fig. 9.11: Averaged performance decay index $e_J(T)$ (–∗–) and maximal performance deviation $e_J^{\max}(T)$ (–▽–) as defined in (9.3) for example (6.16) as a function of the prediction horizon T.

$i = 1, \ldots, N_{x_0}$, as initial state the closed-loop trajectory is computed with the CITOC solution and the suboptimal CFTOC solution for different prediction horizons T, respectively. The performance decay index and the maximum performance deviation are defined as

$$e_J(T) := \frac{1}{N_{x_0}} \sum_{i=1}^{N_{x_0}} \frac{|J_T^*(x_{0,i}) - J_\infty^*(x_{0,i})|}{J_\infty^*(x_{0,i})}, \qquad (9.3a)$$

$$e_J^{\max}(T) := \max_{x_{0,i}} \frac{|J_T^*(x_{0,i}) - J_\infty^*(x_{0,i})|}{J_\infty^*(x_{0,i})}. \qquad (9.3b)$$

One can see from Figure 9.11, where the number of grid points is $N_{x_0} = 1660$, that already for $T = 3$ the (averaged) relative error in the cost is less than 1.5 %. And for increasing prediction horizon the error vanishes quickly.

9.4.3 Constrained PWA System

Consider the piecewise affine system [MR03]

$$\begin{cases} x(t+1) = \begin{cases} \begin{bmatrix} 1 & 1/5 \\ 0 & 1 \end{bmatrix} x(t) + \begin{bmatrix} 0 \\ 1 \end{bmatrix} u(t) + \mathbb{0}_2, & \text{if } [1,\ 0]x(t) \leq 1, \\ \begin{bmatrix} 1/2 & 1/5 \\ 0 & 1 \end{bmatrix} x(t) + \begin{bmatrix} 0 \\ 1 \end{bmatrix} u(t) + \begin{bmatrix} 1/2 \\ 0 \end{bmatrix}, & \text{if } [1,\ 0]x(t) > 1, \end{cases} \\ \begin{bmatrix} -1 & 1 \\ -3 & -1 \\ 1/5 & 1 \\ -1 & 0 \\ 1 & 0 \\ 0 & -1 \end{bmatrix} x(t) \leq \begin{bmatrix} 15 \\ 25 \\ 9 \\ 6 \\ 8 \\ 10 \end{bmatrix}, \quad \text{and} \quad u(t) \in [-1,\ 1]. \end{cases} \qquad (9.4)$$

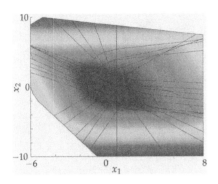

Fig. 9.12: Cost $J_{RH}(x)$ for example (9.4) with $T = 4$. Same color corresponds to the same cost value.

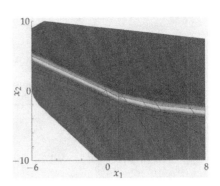

Fig. 9.13: Control law $\mu_{RH}(x)$ for example (9.4) with $T = 4$. Blue ($u = -1$) and red-brown ($u = 1$) marked regions correspond to the saturated control action.

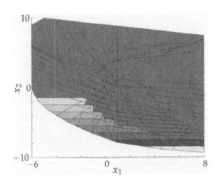

Fig. 9.14: Lyapunov stability region $\mathcal{V}$ (red) and region of attraction $\mathcal{A}$ (red+grey-shades) for example (9.4) with $T = 4$.

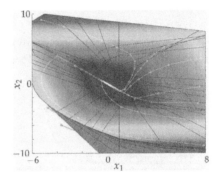

Fig. 9.15: Maximal positively invariant set $\mathcal{O}_{max}$ for example (9.4) with $T = 4$. The same color corresponds to the same cost value. Green marked regions lead to infeasibility.

The constrained finite time optimal control problem (6.1a)–(6.1b) is solved with $Q = I_2$, $R = 1/10$, $P = 0_{2 \times 2}$, and $\mathcal{X}^f = [-10, 10] \times [-10, 10]$ for $p = \infty$. $T = 4$ was chosen as prediction horizon. The results are reported in Figures 9.12–9.15 and in Table 9.3.

For $T = 4$ the region of attraction $\mathcal{A}$ and the maximal positively invariant set $\mathcal{O}_{max}$ are identical[3], cf. Figure 9.14 and Figure 9.15. This means that no limit cycle or stationary points other than the origin exists. Furthermore, the stability region is a strict subset of the feasible region, i.e. there exist regions of the CFTOC solution for $T = 4$ which lead to infeasibility when the receding horizon policy is applied (green marked regions in Figure 9.15).

---

[3] The solution for $T = 3$ was not computed nor tested. The mentioned results for $T = 4$ might also hold for smaller prediction horizons.

9.4 Examples

computation for	T = 4
Lyapunov stability region $\mathcal{V}$	7 iterations
	201 regions
region of attraction $\mathcal{A}$	9 iterations
	370 regions
	$\mathcal{A} = \mathcal{O}_{max} \subset \mathcal{X}$
max. positively invariant set $\mathcal{O}_{max}$	8 iteration
	123 regions
	$\mathcal{A} = \mathcal{O}_{max} \subset \mathcal{X}$
CFTOC solution	119 regions

Tab. 9.3: Computational results for example (9.4).

9.4.4 3-Dimensional Constrained PWA System

Consider the piecewise affine system [MR03]

$$x(t+1) = \begin{cases} \begin{bmatrix} 1 & 1/2 & 3/10 \\ 0 & 1 & 1 \\ 0 & 0 & 1 \end{bmatrix} x(t) + \begin{bmatrix} 0 \\ 0 \\ 1 \end{bmatrix} u(t) + 0_3, & \text{if } x_2(t) \leq 1, \\ \begin{bmatrix} 1 & 1/5 & 3/10 \\ 0 & 1/2 & 1 \\ 0 & 0 & 1 \end{bmatrix} x(t) + \begin{bmatrix} 0 \\ 0 \\ 1 \end{bmatrix} u(t) + \begin{bmatrix} 3/10 \\ 1/2 \\ 0 \end{bmatrix}, & \text{if } x_2(t) > 1, \end{cases}$$

$$\begin{bmatrix} -1 & 0 & 0 \\ 1 & 0 & 0 \\ 0 & -1 & 0 \\ 0 & 1 & 0 \\ 0 & 0 & -1 \\ 0 & 0 & 1 \end{bmatrix} x(t) \leq \begin{bmatrix} 10 \\ 10 \\ 5 \\ 5 \\ 10 \\ 10 \end{bmatrix}, \quad \text{and} \quad u(t) \in [-1, 1].$$

(9.5)

The constrained finite time optimal control problem (6.1a)–(6.1b) is solved with $Q = I_3$, $R = 1/10$, $P = 0_{3\times3}$, and $\mathcal{X}^f = [-10, 10] \times [-10, 10] \times [-10, 10]$ for

computation for	T = 3	T = 4
Lyapunov stability region $\mathcal{V}$	11 iterations	6 iterations
	2 526 regions	2 119 regions
region of attraction $\mathcal{A}$	computation was not possible due to limited amount of memory	7 iterations
		9 969 regions
max. positively invariant set $\mathcal{O}_{max}$	14 iteration	9 iterations
	1 139 regions	1 063 regions
	2 632 infeasible regions	665 infeasible regions
CFTOC solution	329 regions	853 regions

Tab. 9.4: Computation results for example (9.5).

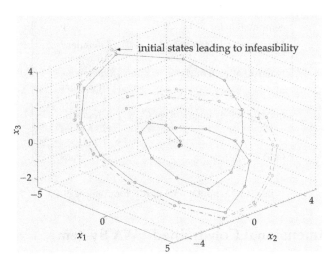

Fig. 9.16: Simulation for example (9.5) with $T = 3$. Red marked trajectories lead to infeasibility. The purple star marks the origin.

$p = \infty$. $T = 3$ (resp. $T = 4$) was chosen as prediction horizon. The results are reported in Table 9.4. There one sees that during the computation of the maximal positively invariant set 2 632 (resp. 665) regions of the CFTOC solution for $T = 3$ (resp. $T = 4$) lead to infeasibility when the receding horizon policy is applied (red marked trajectories in Figure 9.16). This implies that the region of attraction $\mathcal{A} \subseteq \mathcal{O}_{\max} \subset \mathcal{X}$.

10
Stability Tubes

In this chapter the concept of the stability tube is presented. By a posteriori analyzing a nominal stabilizing controller and the corresponding Lyapunov function of a general closed-loop nonlinear system one can compute a set, called the stability tube, in the augmented state-input space. Any control chosen from this set guarantees closed-loop stability and a pre-specified level of closed-loop performance. Furthermore, these stability tubes can serve, for example, as a robustness analysis tool for the given nominal controller as well as a means to obtain an approximation of the nominal control law with lower complexity while preserving closed-loop stability and a certain performance level. For the piecewise affine systems and piecewise affine functions, considered in this manuscript, the overall computation of the stability tubes reduces to basic polytope manipulations.

10.1 Introduction

One of the most challenging tasks after obtaining a controller, based on the nominal model of the real system, is to actually implement this controller (or a simplified approximation of it) in the real process and to give some guarantees about the closed-loop stability or performance when uncertainty in the system, measurement noise, or actuator uncertainty is present.

Traditionally, a (simplified) nominal controller is extensively tested for different operating points on the real process or on a simulation model, to show if the specifications are fulfilled. However, this procedure is highly time and cost consuming and strong guarantees on the closed-loop stability and/or performance for the whole state space are often neglected.

In the presence of system and/or input constraints, a large number of robust control methods are based on the *min-max model predictive control* idea, cf. for example [MRRS00, BM99b, FIAF03] and the references therein. That is, given a model including a specific uncertainty or noise description, the aim is to minimize or counter-act over a finite prediction horizon a possible worst case scenario coming from the uncertainty acting on the system. Unfortunately, min-max based techniques are (even though theoretically very interesting) notoriously difficult to solve, computationally highly demanding, and in many cases even prohibitive. Furthermore, as soon as an approximation of the optimal min-max solution for implementation purposes is computed, the theoretical stability and performance guarantees might be lost. In recent years, a whole plethora of other robust synthesis techniques for systems with constraints have appeared; two examples are [RM05, MRVK06].

Here we take a different approach, where the inherent robust stability property of a closed-form nominal control law is investigated and the concept of the *stability tube* is introduced. By a posteriori analyzing a nominal closed-form stabilizing controller and corresponding Lyapunov function of the general closed-loop nonlinear system one can compute a set, called a stability tube, in the augmented state-input space. Any controller chosen from this set guarantees closed-loop stability and a pre-specified level of closed-loop performance. These stability tubes can serve, for example, as a *robustness analysis* tool for the given nominal controller as well as a means to obtain an approximation of the nominal control law with lower complexity while preserving closed-loop stability and a certain performance level. For the piecewise affine systems and piecewise affine functions considered in this manuscript, the overall computation of the stability tube reduces to basic polytope manipulations.

10.2 Constrained Finite Time Optimal Control of Piecewise Affine Systems

Without repeating all the detail of Part II, a brief overview about an interesting problem class, for which a stability tube computation is feasible and 'simple', is given in the following.

Consider the class of discrete-time, stabilizable, linear hybrid systems that can be described as constrained *piecewise affine* (PWA) systems of the following form

10.2 Constrained Finite Time Optimal Control of PWA Systems

$$x(t+1) = f_{\text{PWA}}(x(t), u(t))$$
$$:= A_d x(t) + B_d u(t) + a_d, \quad \text{if } \begin{bmatrix} x(t) \\ u(t) \end{bmatrix} \in \mathcal{D}_d, \quad (4.1)$$

where $t \geq 0$, the domain $\mathcal{D} := \cup_{d=1}^{N_\mathcal{D}} \mathcal{D}_d$ of $f_{\text{PWA}}(\cdot, \cdot)$ is a non-empty compact set in $\mathbb{R}^{n_x + n_u}$ with $N_\mathcal{D} < \infty$ the number of system dynamics, and $\{\mathcal{D}_d\}_{d=1}^{N_\mathcal{D}}$ denotes a polyhedral partition of the domain $\mathcal{D}$, where the closure of $\mathcal{D}_d$ is given by $\bar{\mathcal{D}}_d := \{[\begin{smallmatrix} x \\ u \end{smallmatrix}] \in \mathbb{R}^{n_x+n_u} \mid D_d^x x + D_d^u u \leq D_d^0\}$ and $\text{int}(\mathcal{D}_d) \cap \text{int}(\mathcal{D}_j) = \emptyset$, for all $d \neq j$.

Linear state constraints ($x \in \mathbb{X}$) and input constraints ($u \in \mathbb{U}$) of the general form $C^x x + C^u u \leq C^0$ are naturally incorporated in the description of $\mathcal{D}_d$.

Furthermore, let us define for the aforementioned piecewise affine system (4.1) the *constrained finite time optimal control* (CFTOC) problem

$$J_T^*(x(0)) := \min_{U_T} J_T(x(0), U_T) \quad (6.1a)$$

$$\text{subj. to } \begin{cases} x(t+1) = f_{\text{PWA}}(x(t), u(t)) \\ x(T) \in \mathcal{X}^f, \end{cases} \quad (6.1b)$$

where

$$J_T(x(0), U_T) := \ell_T(x(T)) + \sum_{t=0}^{T-1} \ell(x(t), u(t)) \quad (6.1c)$$

is the cost function (also called performance index), $\ell(\cdot, \cdot)$ the stage cost, $\ell_T(\cdot)$ the final penalty function, $U_T := \{u(t)\}_{t=0}^{T-1}$ is the optimization variable, $T < \infty$ is the prediction horizon, and $\mathcal{X}^f$ is a compact terminal target set in $\mathbb{R}^{n_x}$.

And consider the two following restrictions to the CFTOC problem (6.1a)–(6.1c)

Problem 6.1 (PWA system, 1-/∞-norm based cost). Let the cost function (6.1c) be composed of

$$\ell(x(t), u(t)) := \|Qx(t)\|_p + \|Ru(t)\|_p \quad \text{and}$$
$$\ell_T(x(T)) := \|Px(T)\|_p, \quad \square$$

where $\|\cdot\|_p$ with $p \in \{1, \infty\}$ denotes the standard vector 1-/∞-norm [HJ85], and

Problem 10.1 (Constrained LTI system, quadratic cost).

$$f_{\text{PWA}}(x(t), u(t)) := Ax(t) + Bu(t), \quad \text{if } \begin{bmatrix} x(t) \\ u(t) \end{bmatrix} \in \mathcal{D},$$
$$\ell(x(t), u(t)) := x(t)' Q x(t) + u(t)' R u(t), \quad \text{and}$$
$$\ell_T(x(T)) := x(T)' P x(T). \quad \square$$

In both CFTOC Problem 6.1 and Problem 10.1, the solution is a time-varying piecewise affine state feedback control law defined over a polyhedral partition, which is stated in the following theorem, confer also Chapter 6 and e.g. [Bor03].

Theorem 6.5 (Solution to CFTOC). The solution to the optimal control problem (6.1a)–(6.1b), restricted to Problem 6.1 or 10.1, is a time-varying piecewise affine function of the initial state $x(0)$

$$\mu(x(0), t) = K_{T-t,i} x(0) + L_{T-t,i}, \quad \text{if} \quad x(0) \in \mathcal{P}_i \qquad (6.5)$$

with $u^*(t) = \mu(x(0), t)$, where $t = 0, \ldots, T-1$ and $\{\mathcal{P}_i\}_{i=1}^{N_P}$ is a polyhedral partition of the set of feasible states $x(0)$

$$\mathcal{X}_T = \cup_{i=1}^{N_P} \mathcal{P}_i,$$

with the closure of $\mathcal{P}_i$ given by $\bar{\mathcal{P}}_i = \{x \in \mathbb{R}^{n_x} \mid P_i^x x \leq P_i^0\}$. ∎

In the case that a *receding horizon* (RH) control policy is used in closed-loop, the control is given as a time-invariant state feedback control law of the form

$$\mu_{\text{RH}}(x(t)) := \mu(x(t), 0) = K_{T,i} x(t) + L_{T,i}, \quad \text{if} \quad x(t) \in \mathcal{P}_i, \qquad (6.14)$$

where $i = 1, \ldots, N_P$ and $u(t) = \mu_{\text{RH}}(x(t))$ for $t \geq 0$.

Assumption 10.2 (Stability, all time feasibility). Note that throughout this chapter it is assumed that the parameters T, Q, R, P, and $\mathcal{X}^f$ of the CFTOC Problem 6.1 or 10.1 are chosen in such a way that (6.14) is closed-loop stabilizing, feasible for all time (cf. Definition 3.4), and that a polyhedral piecewise affine Lyapunov function of the form

$$V(x) = V_i^x x + V_i^0, \quad \text{if} \quad x \in \mathcal{P}_i,$$

where $i = 1, \ldots, N_P$, for the *closed-loop system*

$$x(t+1) = f^{\text{CL}}(x(t)) := f_{\text{PWA}}(x(t), \mu_{\text{RH}}(x(t))), \qquad (10.1)$$

with $x(t) \in \mathcal{X}_T$, exists and is given. □

This is usually not a restricting requirement but rather the aim of most (if not all) control strategies. While it is in general a challenging problem to compute Lyapunov functions for nonlinear systems, a construction of Lyapunov functions for constrained piecewise affine systems has been shown e.g. in [LHW+05, BGLM05]. Furthermore note, that if the parameters are chosen according to e.g. Theorem 3.5 one can simply take $V(\cdot) = J_T^*(\cdot)$. A possible and indeed obvious choice is to employ the value function $J_\infty^*(\cdot)$ from the corresponding infinite time solution as Lyapunov function, cf. Chapter 7.

In the course of this chapter the focus lies on the derivation of a tool for the robustness analysis as well as approximation and complexity reduction of the control law $\mu_{RH}(\cdot)$ without losing closed-loop stability nor feasibility for all time, while preserving a certain level of optimality. Therefore, the following sections introduce the set concept of a *stability tube* of a given nominal controller.

10.3 Stability Tubes for Nonlinear Systems

Consider the (possibly discontinuous) discrete-time nonlinear system

$$x(t+1) = f(x(t), u(t)), \quad \text{where} \quad \begin{bmatrix} x(t) \\ u(t) \end{bmatrix} \in \mathcal{D}, \quad (10.2a)$$

with domain[1] $\mathcal{D} \subseteq \mathbb{R}^{n_x+n_u}$, $\mathbb{0}_{n_x} = f(\mathbb{0}_{n_x}, \mathbb{0}_{n_u})$, and a corresponding nominal stabilizing control law $u(t) = \mu(x(t))$. The stable closed-loop autonomous system with equilibrium point $x_{eq} = \mathbb{0}_{n_x}$ is then given by

$$x(t+1) = f^{CL}(x(t)) := f(x(t), \mu(x(t))), \quad \text{with} \quad x(t) \in \mathcal{X} \quad (10.2b)$$

and $\mathcal{X} \subseteq \mathbb{R}^{n_x}$ being positively invariant.

The underlying core idea presented here, lies in the inherent freedom of the Lyapunov decay inequality (2.3c) of Theorem 2.11, repeated in the following from Chapter 2 for completeness.

Theorem 2.11 (Asymptotic/exponential stability). Let $\mathcal{X} \subseteq \mathbb{X}$ be a bounded positively invariant set for the autonomous system (10.2b) that contains a neighborhood $\mathcal{N}(x_{ep})$ of the equilibrium $x_{ep} = \mathbb{0}_{n_x}$, and let $\underline{\alpha}(\cdot)$, $\bar{\alpha}(\cdot)$, and $\beta(\cdot)$ be K-class functions, cf. Definition 1.44.

If there exists a non-negative function $V : \mathcal{X} \to \mathbb{R}_{\geq 0}$ with $V(\mathbb{0}_{n_x}) = 0$ such that

$$V(x) \geq \underline{\alpha}(\|x\|), \quad x \in \mathcal{X}, \quad (2.3a)$$
$$V(x) \leq \bar{\alpha}(\|x\|), \quad x \in \mathcal{N}(x_{ep}), \quad (2.3b)$$
$$\Delta V(x) := V(f^{CL}(x)) - V(x) \leq -\beta(\|x\|), \quad x \in \mathcal{X}, \quad (2.3c)$$

then the following results holds:

(a) The equilibrium point $\mathbb{0}_{n_x}$ is *asymptotically stable* in the Lyapunov sense in $\mathcal{X}$.

---
[1] Note that standard state constraints ($x \in \mathbb{X} \subseteq \mathbb{R}^{n_x}$) and input constraints ($u \in \mathbb{U} \subseteq \mathbb{R}^{n_u}$) are included in the general formulation of $\mathcal{D}$.

(b) If $\underline{\alpha}(\|x\|) := \underline{a}\|x\|^\gamma$, $\overline{\alpha}(\|x\|) := \overline{a}\|x\|^\gamma$, and $\beta(\|x\|) := b\|x\|^\gamma$ for some positive constants $\underline{a}, \overline{a}, b, \gamma > 0$ then the equilibrium point $\mathbb{0}_{n_x}$ is *locally exponentially stable*. Moreover, if the inequality (2.3b) holds for $\mathcal{N}(x_{\mathrm{ep}}) = \mathcal{X}$, then the equilibrium point $\mathbb{0}_{n_x}$ is *exponentially stable* in the Lyapunov sense in $\mathcal{X}$. ∎

Simply speaking, if all the prerequisites of Theorem 2.11 are fulfilled with a given nominal controller $\mu(\cdot)$, the resulting behavior of the closed-loop system is stabilizing. If, for the given function $V(\cdot)$, $\beta(\cdot)$ is now relaxed and a controller $\widetilde{\mu}(\cdot)$ is unknown, one can (possibly) find a whole set of controllers $\widetilde{\mu}(\cdot)$ that will render the closed-loop system stabilizing and feasible, i.e. for which the function $V(\cdot)$ is also a (control) Lyapunov function. (Note, that setting $\beta(\cdot)$ close to the zero-function, i.e. for example $V(f^{\mathrm{CL}}(x)) - V(x) \leq -\beta\|x\|$ with $0 < \beta \ll 1$ is sufficient for pure asymptotic stability.) These sets are denoted in the following as *stability tubes*.

Definition 10.3 (Stability tube). Let $V(\cdot)$ be a Lyapunov function for the nonlinear closed-loop system (10.2b) with $x(t) \in \mathcal{X}$ under a stabilizing nominal controller $u(t) = \mu(x(t))$ and let the prerequisites of Theorem 2.11 be fulfilled. Furthermore, let $\beta(\cdot)$ be a K-class function. Then the set

$$\mathcal{S}(V,\beta) := \left\{ \begin{bmatrix} x \\ u \end{bmatrix} \subseteq \mathbb{R}^{n_x \times n_u} \;\middle|\; x \in \mathcal{X},\, f(x,u) \in \mathcal{X},\, \begin{bmatrix} x \\ u \end{bmatrix} \in \mathcal{D}, \right.$$
$$\left. V(f(x,u)) - V(x) \leq -\beta(\|x\|) \right\}$$

is called a *stability tube*. □

Theorem 10.4 (Stability tube). Let the assumptions of Definition 10.3 be fulfilled. Then every control law $u(t) = \widetilde{\mu}(x(t))$ with $x(t) \in \mathcal{X}$ (also *any* sequence of control samples $u(t)$) fulfilling

$$\begin{bmatrix} x(t) \\ u(t) \end{bmatrix} \in \mathcal{S}(V,\beta)$$

asymptotically stabilizes the system (10.2a) to the origin $\mathbb{0}_{n_x}$, where $x(t) \in \mathcal{X}$. □

Proof. Theorem 10.4 directly follows from $V(\cdot)$ being a Lyapunov function for the controlled system and Theorem 2.11(a). ∎

Naturally, for general nonlinear systems, the stability tube $\mathcal{S}(V,\beta)$ can basically take almost any form. Note, however, that for the considered class of piecewise affine systems, piecewise affine control laws, and piecewise affine Lyapunov functions with $\beta(\cdot)$ consisting of a sum of weighted vector 1-/∞-norms, the stability tube $\mathcal{S}(V,\beta)$, as will be indicated in Section 10.4, can be described by a collection of polytopic sets in the state-input space and can be computed with basic polytopic operations. Moreover, for piecewise affine Lyapunov functions exponential stability can be guaranteed.

Theorem 10.5 (Stability tube with PWA Lyapunov function). Let the assumptions of Definition 10.3 be fulfilled, $\mathcal{X}$ be compact, and the Lyapunov function is of the form

$$V(x) = V_i^x x + V_i^0, \quad \text{if } x \in \mathcal{X}_i, \tag{10.4}$$

where $i = 1, \ldots, N_\mathcal{X}$, $N_\mathcal{X} < \infty$, $\mathcal{X} = \bigcup_{i=1}^{N_\mathcal{X}} \mathcal{X}_i$, the closure of $\mathcal{X}_i$ is given by $\bar{\mathcal{X}}_i := \{x \in \mathbb{R}^{n_x} \mid X_i^x x \leq X_i^0\}$, and $\text{int}(\mathcal{X}_i) \cap \text{int}(\mathcal{X}_j) = \emptyset$, for all $i \neq j$.

Then every control law $u(t) = \tilde{\mu}(x(t))$ with $x(t) \in \mathcal{X}$ (also *any* sequence of control samples $u(t)$) fulfilling

$$\begin{bmatrix} x(t) \\ u(t) \end{bmatrix} \in \mathcal{S}(V, \beta)$$

exponentially stabilizes the system (10.2a) to the origin 0_{n_x}, where $x(t) \in \mathcal{X}$. □

Proof. Theorem 10.4 directly follows from $V(\cdot)$ being a Lyapunov function for the controlled system, Theorem 2.11(b), and the fact that for a Lyapunov function of the form (10.4) there always exist $\underline{a}, \bar{a}, b > 0$ with $\gamma = 1$. ∎

Note that due to the sufficiency (and thus possible inherent conservatism) of the Lyapunov stability criterion, the stability tube as in Definition 10.3 is not unique for a specific system nor is it the maximal possible stability tube, but depends on the Lyapunov function $V(\cdot)$ used and corresponding function $\beta(\cdot)$.

To demonstrate the idea of the stability tube, consider the following simple piecewise affine example:

Example 10.6 (Stability tube).

$$x(t+1) = \begin{cases} \frac{4}{5} x(t) + 2u(t), & \text{if } x > 0, \\ -\frac{6}{5} x(t) + u(t), & \text{if } x \leq 0, \end{cases} \tag{10.5a}$$

$$\text{with } u(t) \in [-1, 1] \quad \text{and} \quad x(t) \in [-2, 2]. \tag{10.5b}$$

When solving the CFTOC Problem 6.1 with $p = 1$ and $Q = 1$, $R = 1$, $P = 0$, $T = 5$ one obtains the result illustrated in Figure 10.1. The optimal receding horizon control law $\mu_{\text{RH}}(\cdot)$ (dashed line) is a PWA function defined over $N_P = 3$ regions, while the stability tube $\mathcal{S}(J_5^*, \beta\|x\|_1)$ for the region is shaded. Note that $\mathcal{S}(J_5^*, \beta\|x\|_1)$ is represented by a collection of also 3 polytopes and that $J_5^*(\cdot)$ is a Lyapunov function for the closed-loop system.

Moreover, with the choice of $\beta(\cdot)$ a detuning of the *closed-loop performance*

$$J_\infty\big(x(0), \{u(t)\}_{t=0}^\infty\big) := \lim_{T \to \infty} \sum_{t=0}^T \ell(x(t), u(t)),$$

with some control law $u(t) = \tilde{\mu}(x(t))$, compared to the nominal control solution $\mu(\cdot)$, can be performed. Thus, one can try to find an approximation $\tilde{\mu}(\cdot)$

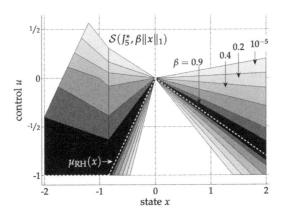

Fig. 10.1: Optimal control law $\mu_{\text{RH}}(\cdot)$ (dashed line) for Example 10.6. The shaded sets, comprising 3 polytopes, are the corresponding stability tubes $\mathcal{S}(J_5^*, \beta\|x\|_1)$ for different values of β. With increasing β, $\mathcal{S}(\cdot, \cdot)$ are subsets of each other.

'inside' the stability tube without losing closed-loop stability, all time feasibility, while still guaranteeing a given, bounded performance decay of $\eta\,\%$. The influence of a different $\beta(\cdot)$ is elaborated in the following.

Corollary 10.7 (Performance bound). Let the assumptions of Definition 10.3 be fulfilled, the stage cost $\ell(\cdot, \cdot)$ be lower bounded by some K-class function, and $\beta > 0$. Then every control law $u(t) = \widetilde{\mu}(x(t))$ with $x(t) \in \mathcal{X}$ (also *any* sequence of control samples $u(t)$) fulfilling

$$\begin{bmatrix} x(t) \\ u(t) \end{bmatrix} \in \mathcal{S}(V, \beta \ell) \tag{10.6}$$

asymptotically[2] *stabilizes* the system (10.2a), where $x(t) \in \mathcal{X}$, to the origin $\mathbb{0}_{n_x}$ and guarantees a level of *closed-loop performance* given by

$$\lim_{T \to \infty} \sum_{t=0}^{T} \ell(x(t), \widetilde{\mu}(x(t))) \leq \frac{1}{\beta} V(x(0)). \tag{10.7}$$

□

Proof. In addition to Theorem 10.4 we have from the definition of $\mathcal{S}(V, \beta \ell)$ that $u(t) = \widetilde{\mu}(x(t))$ fulfills $V(f(x, \widetilde{\mu}(x))) - V(x) \leq -\beta \ell(x, \widetilde{\mu}(x))$. This implies

$$\lim_{T \to \infty} \sum_{t=0}^{T} \ell(x(t), \widetilde{\mu}(x(t))) \leq \frac{1}{\beta} \lim_{T \to \infty} \sum_{t=0}^{\infty} V(x(t)) - V(f(x(t), \widetilde{\mu}(x(t))))$$

$$= \frac{1}{\beta} \left(V(x(0)) - \lim_{t \to \infty} V(x(t)) \right) = \frac{1}{\beta} V(x(0)). \blacksquare$$

---

[2] If $V(\cdot)$ is of the piecewise affine form (10.4) and $\mathcal{X}$ is compact, we have from Theorem 10.5 that $\{u(t)\}_{t=0}^{\infty}$ *exponentially stabilizes* the system to the origin.

Example 10.6(cont.) (Stability tube with performance). Figure 10.1 illustrates for Example 10.6 the stability tubes for a variety of different β and fixed $V(\bullet) = J_5^*(\bullet)$. Note, that (naturally) the stability tubes are subsets of each other and collapse to the control law $\mu_{\text{RH}}(\bullet)$ itself as $\beta \to 1$.

From (10.7) one can define the *(relative) performance decay* η [in %] with respect to $V(x(0))$

$$\eta := \frac{|J_\infty(x(0), \widetilde{\mu}(\bullet)) - V(x(0))|}{V(x(0))} \cdot 100,$$

which is related to $\beta > 0$ via

$$\beta(\eta) = \left(1 + \frac{\eta \ [\text{in \%}]}{100}\right)^{-1}.$$

In the case that the constrained infinite time optimal control (CITOC) problem is solved, i.e. (roughly speaking) considering the CFTOC problem (6.1a)–(6.1c) with $T \to \infty$, one obtains *the* optimal solution $\mu_\infty^*(\bullet)$ with corresponding value function $J_\infty^*(\bullet)$, cf. Chapter 7. It is proved in Chapter 7 that $J_\infty^*(\bullet)$ is a Lyapunov function for the closed-loop system.

Corollary 10.8 (Performance bound with J_∞^*). Let the assumptions of Corollary 10.7 be fulfilled and $V(\bullet) = J_\infty^*(\bullet)$. Then every control law $u(t) = \widetilde{\mu}(x(t))$ with $x(t) \in \mathcal{X}$ (also *any* sequence of control samples $u(t)$) fulfilling

$$\begin{bmatrix} x(t) \\ u(t) \end{bmatrix} \in \mathcal{S}(J_\infty^*, \beta \ell) \tag{10.8}$$

asymptotically[2] *stabilizes* the system (10.2a), where $x(t) \in \mathcal{X}$, to the origin 0_{n_x} and guarantees a level of *closed-loop performance* given by

$$J_\infty^*(x(0)) \leq \lim_{T \to \infty} \sum_{t=0}^{T} \ell(x(t), \widetilde{\mu}(x(t))) \leq \frac{1}{\beta} J_\infty^*(x(0)), \tag{10.9}$$

and $0 < \beta \leq 1$. □

Proof. In addition to Corollary 10.7 we have

$$\lim_{T \to \infty} \sum_{t=0}^{T} \ell(x(t), \widetilde{\mu}(x(t))) \geq \min_{u(t) \in \mathbb{U}} \lim_{T \to \infty} \sum_{t=0}^{T} \ell(x(t), u(t)) =: J_\infty^*(x(0)),$$

and $J_\infty^*(x) \leq \frac{1}{\beta} J_\infty^*(x)$ implies $0 < \beta \leq 1$. ∎

10.4 Computation of Stability Tubes for Piecewise Affine Systems

As mentioned above, for the class of piecewise affine systems (4.1), corresponding piecewise affine Lyapunov function $V(\bullet)$, and $\beta(\bullet)$ consisting of a sum of weighted linear vector norms, the stability tube can be represented and 'easily' computed as a collection (or union) of polytopes of the form

$$\mathcal{S}(V, \beta) := \cup_{l=1}^{N_\mathcal{S}} \mathcal{S}_l, \qquad (10.10)$$

where the closure of $\mathcal{S}_l$ is

$$\bar{\mathcal{S}}_l := \left\{ \begin{bmatrix} x \\ u \end{bmatrix} \in \mathbb{R}^{n_x+n_u} \mid S_l^{xu} \begin{bmatrix} x \\ u \end{bmatrix} \leq S_l^0 \right\},$$

cf. for example Figure 10.1. Without going into technical detail, the polytope $\mathcal{S}_l$ in the state-input space, with $l = 1, \ldots, N_\mathcal{S}$, can be obtained from Definition 10.3 itself, i.e.

$$x \in \mathcal{P}_i, \qquad (10.11a)$$
$$f_{\text{PWA}}(x, u) \in \mathcal{P}_j, \qquad (10.11b)$$
$$\begin{bmatrix} x \\ u \end{bmatrix} \in \mathcal{D}, \qquad (10.11c)$$
$$V(f_{\text{PWA}}(x, u)) - V(x) \leq -\beta(\|x\|_p) \qquad (10.11d)$$

for some $i, j = 1, \ldots, N_\mathcal{P}$. Note that a large number of transitions of states from region $\mathcal{P}_i$ to $\mathcal{P}_j$ can usually be out-ruled by $V(f(x, u)) < V(x)$. Depending on the form of $\beta(\cdot)$ an additional projection onto the state-input space might be necessary.

By construction, we have the following properties: (a) for some index set $\mathcal{I}_i \subseteq \{1, \ldots N_\mathcal{S}\}$, the union $\cup_{l \in \mathcal{I}_i} \mathcal{S}_l$ is defined over the region $\mathcal{P}_i$ of the controller, and (b), $\sum_{i=1}^{N_\mathcal{P}} |\mathcal{I}_i| = N_\mathcal{S}$. This means that each $\mathcal{S}_l$ is defined over a single region $\mathcal{P}_i$, i.e. if for some i_1 and l we have $\text{proj}_x(\mathcal{S}_l) \subseteq \mathcal{P}_{i_1}$ then there does not exists a $i_2 \neq i_1$ with $\text{proj}_x(\mathcal{S}_l) \subseteq \mathcal{P}_{i_2}$. (We remark, that simulations seem to indicate that most often $\mathcal{I}_i = 1$ for all i, i.e. only one $\mathcal{S}_l$ is defined over $\mathcal{P}_i$.)

A reduction of the complexity $N_\mathcal{S}$ by, for example, merging neighboring polytopes is, in the author's experience, most often very successful.

10.5 Comments on Stability Tubes

As indicated before, the stability tube $\mathcal{S}(V, \beta)$ can, for example, be used for the purpose of robustness analysis or complexity reduction of a nominal controller.

Robustness Analysis

The computation of robust control laws for constrained piecewise affine systems is in general a hard problem. The stability tube allows a direct (possibly conservative) *analysis* of the inherent robust stability and robust performance

10.5 COMMENTS ON STABILITY TUBES

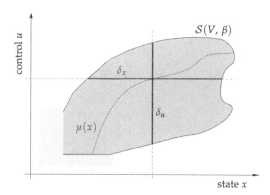

Fig. 10.2: Allowable uncertainties δ_u in the actuator or noise δ_x of the state measurement without the loss of closed-loop stability when the nominal controller $\mu(x)$ is applied to the system.

property of the nominal controller $u(t) = \mu(x(t))$ with respect to *sensor measurement noise* δ_x and/or *actuator uncertainty* δ_u, cf. Figure 10.2.

From Theorem 10.4 follows that if for all $t \geq 0$, $\delta_x(t)$ and $\delta_u(t)$ are such that

$$\begin{bmatrix} x(t)+\delta_x(t) \\ \mu(x(t)+\delta_x(t))+\delta_u(t) \end{bmatrix} \in \mathcal{S}(V, \beta),$$

then the closed-loop controlled system is stabilizing. More specifically, the allowable uncertainty of the actuator, i.e. $\delta_u \in \Delta_u$, can be explicitly computed via

$$\Delta_u(x) := \left\{ \delta_u \in \mathbb{R}^{n_u} \;\middle|\; \begin{bmatrix} x \\ \mu(x)+\delta_u \end{bmatrix} \in \mathcal{S}(V, \beta) \right\},$$

cf. the blue solid line in Figure 10.2. In an analogous way, the effect of the allowable noisy state measurement $x(t) + \delta_x(t)$ can be determined.

Note that for the considered piecewise affine systems, piecewise control laws and polytopic stability tube, all the operations are basic operations on polytopes.

Furthermore, in the case of *additive state disturbances* $w(t) \in \mathcal{W}$ on the piecewise affine systems, i.e.

$$x(t+1) = f_{\text{PWA}}(x(t), u(t)) + w(t),$$

one can approximate some $\mathcal{W}$ from Δ_u.

Controller Complexity Reduction and Approximation

As the system dimension n_x and control dimension n_u are fixed, the storage demand (or complexity) of a closed-form control law (6.14) is influenced solely by the defining polyhedral partition, i.e. the number $N_\mathcal{P}$ of defining state space polyhedral regions $\mathcal{P}_i$ and the number of their respective facets.

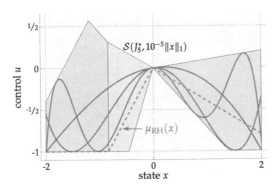

Fig. 10.3: Stabilizing polynomial approximations (solid lines) of the control law $\mu_{RH}(\cdot)$ (dashed red line) 'inside' the stability tube $\mathcal{S}(J_5^*, 10^{-5}\|x\|_1)$ of Example 10.6.

Unfortunately, depending on the structure and parameters of the underlying system and optimization problem, one of the main drawbacks with optimal closed-form control laws $\mu_{RH}(\cdot)$ is the possible worst case exponential 'explosion' [Bor03, BMDP02] in the number of regions $N_\mathcal{P}$. But even in an average case the number $N_\mathcal{P}$ tends to be very large and above the storage limit of most control devices. Therefore, it is often essential for a real-life implementation of the closed-form solution to find an appropriate approximation of the controller or a controller with reduced complexity.

Several authors recently addressed the complexity reduction or approximation issue by either modifying the original CFTOC problem, retrieving a suboptimal solution of the CFTOC problem, or by post-processing the computed optimal controller, cf. e.g. [BF03, GTM04, Gri04, TJB03]. However, only a few of the approaches in the literature give guarantees on the preservation of closed-loop stability and/or performance.

Here, Theorem 10.4 together with Corollary 10.7 directly implies that *any* approximative controller $\tilde{\mu}(\cdot)$ or controller with reduced complexity 'living' inside the stability tube $\mathcal{S}(V, \beta)$ will stabilize the system with a guaranteed pre-specified bounded performance loss.

The authors in [KCHF07] propose to compute control laws $\tilde{\mu}(\cdot)$ in the form of simple (possibly piecewise) polynomials that 'live' inside of the stability tube to guarantee stability and performance. The computation is based on the aforementioned stability tube and the technique of sum-of-squares [Par04, PL03, PPSP04]. Even though the approach is computationally demanding it serves as a technique to simplify the control law $\mu_{RH}(\cdot)$ drastically and enables one to control small and *very fast* sampled systems. Figure 10.3 depicts possible stabilizing polynomial approximations of the control law $\mu_{RH}(\cdot)$ for Example 10.6.

10.6 Examples

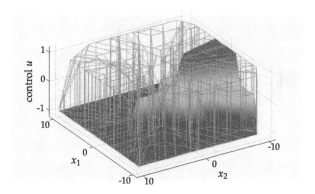

Fig. 10.4: Stability tube $\mathcal{S}(J_\infty^*, 10^{-5}\|x\|_\infty)$ (in green wireframe) together with the optimal control law $\mu_\infty^*(\cdot)$ for the example in Section 10.6.1.

10.6 Examples

10.6.1 Constrained PWA Sine-Cosine System

Consider once more the constrained piecewise affine system [BM99a] from Chapter 6 and 7:

$$\begin{cases} x(t+1) = \frac{4}{5}\begin{bmatrix} \cos\alpha(x(t)) & -\sin\alpha(x(t)) \\ \sin\alpha(x(t)) & \cos\alpha(x(t)) \end{bmatrix} x(t) + \begin{bmatrix} 0 \\ 1 \end{bmatrix} u(t), \\ \alpha(x(t)) = \begin{cases} \frac{\pi}{3} & \text{if } [1\ 0]x(t) \geq 0, \\ -\frac{\pi}{3} & \text{if } [1\ 0]x(t) < 0, \end{cases} \\ x(t) \in [-10, 10] \times [-10, 10], \quad \text{and} \quad u(t) \in [-1, 1]. \end{cases} \quad (6.16)$$

The CITOC Problem 7.1 is solved with $Q = I_2$, $R = 1$, and $\mathcal{X}_0 = [-10, 10] \times [-10, 10]$ for $p = \infty$.

The piecewise affine infinite time control law $u(t) = \mu_\infty^*(x(t))$ and corresponding piecewise affine value function $J_\infty^*(\cdot)$ are defined over $N_P = 252$ polyhedral state space regions. The related stability tube $\mathcal{S}(J_\infty^*, 10^{-5}\|x\|_\infty)$ comprises also $N_\mathcal{S} = 252$ polyhedral regions in the state-input space and is depicted together with $\mu_\infty^*(\cdot)$ in Figure 10.4 and 10.5. The optimal merging technique [GTM04] reduced the complexity $N_\mathcal{S} = 252$ of the stability tube to 110 regions within 17 seconds.

The stability tube for this example was obtained in 133 seconds on a Pentium 4, 2.8 GHz machine on Linux running MATLAB®, the LP solver of NAG [Num02], and MPT [KGB04, KGBM03].

Notice from Figure 10.4 that the optimal control law $\mu_\infty^*(\cdot)$ is inherently 'quite' robustly stable against, for example, uncertainties δ_u in the actuator, i.e. $u(t) = \mu_\infty^*(x(t)) + \delta_u(t)$, or measurement noise δ_x of the system's state,

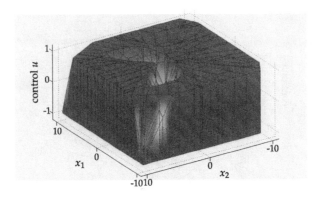

Fig. 10.5: Stability tube $\mathcal{S}(J_\infty^*, 10^{-5}\|x\|_\infty)$ for the example in Section 10.6.1.

i.e. $x_{\text{meas}}(t) = x(t) + \delta_x(t)$ when applying $u(t) = \mu_\infty^*(x_{\text{meas}}(t))$ to the system. Figure 10.6 illustrates this fact, that when bounded random actuator noise $\delta_u(t)$ is acting on the μ_∞^*-controlled system, the closed-loop trajectories are still stabilizing.

10.6.2 Car on a PWA Hill

Recall the example from Section 6.6.2 (page 61ff) in which a car is moving horizontally on a piecewise affine 'environment'. The goal of the car is to climb to the top of a steep hill and then to maintain its position at the top (the origin), without falling from the piecewise affine environment. The discrete-time model is given by the following constrained and discontinuous piecewise affine system

$$x(t+1) = \begin{bmatrix} 1 & 1/2 \\ 0 & 1 \end{bmatrix} x(t) + \begin{bmatrix} 1/8 \\ 1/2 \end{bmatrix} u(t) + a(x(t)), \qquad (6.18a)$$

where

$$a(x(t)) = \begin{cases} 0_2, & \text{if } [1\ 0]x(t) \in [-\tfrac{1}{2},\ 1], \\ -\tfrac{1}{4}g \sin(20\tfrac{\pi}{180}) \begin{bmatrix} 1 \\ 2 \end{bmatrix}, & \text{if } [1\ 0]x(t) \in [-2,\ -\tfrac{1}{2}], \\ 0_2, & \text{if } [1\ 0]x(t) \in [-3,\ -2], \\ -\tfrac{1}{4}g \sin(-5\tfrac{\pi}{180}) \begin{bmatrix} 1 \\ 2 \end{bmatrix}, & \text{if } [1\ 0]x(t) \in [-4,\ -3], \end{cases}$$

g is the gravitational constant, and the control action is constrained by

$$|u(t)| \leq 2 \quad \text{and} \quad |u(t+1) - u(t)| \leq 40.$$

The CFTOC Problem 6.1 was solved with $p = 1$, $Q = \text{diag}([100,\ 1]')$, $R = 5$, $T = 9$, and P and $\mathcal{X}^f$ appropriately chosen such that the closed-loop system under receding control $\mu_{\text{RH}}(\cdot)$ is stabilizing and $J_{\text{RH}}(\cdot)$ is a corresponding

10.6 EXAMPLES

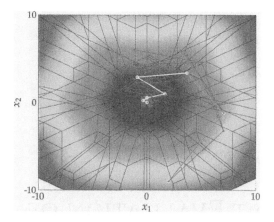

Fig. 10.6: Closed-loop trajectory starting from $x(0) = [2\ -9]'$ using the optimal control law $\mu_\infty^*(\cdot)$ (in white) together with several different stabilizing trajectories (in magenta) also starting at $x(0)$ using random control samples $u(t)$ from the stability tube simulating actuator noise. The coloring scheme of the state space partition indicates the value of the Lyapunov function $J_\infty^*(\cdot)$.

Lyapunov function. $\mu_{\text{RH}}(\cdot)$ and $J_{\text{RH}}(\cdot)$ are defined over 2083 polyhedral state space regions.

The stability tube $\mathcal{S}(J_{\text{RH}}(\cdot), 10^{-5}\|x\|_1)$ comprises 23216 polyhedral regions in 3D and was obtained in 5.7 hours on a Pentium Core 2 Duo, 1.83 GHz machine running MATLAB® 7.3.0, the LP solver of CPLEX [ILO], and MPT 2.6.2 [KGB04, KGBM03]. Note that no simplification of the representation of the stability tube nor any optimization of the computation were pursued.

As indicated in the previous sections, this stability tube can be used for different analysis or synthesis purposes.

11

EFFICIENT EVALUATION OF PIECEWISE CONTROL LAWS DEFINED OVER A LARGE NUMBER OF POLYHEDRA

The on-line evaluation of a piecewise state feedback control law requires the determination of the state space region in which the measured state lies, in order to decide which 'piece' of the piecewise control law to apply. This procedure is called the point location problem, and the rate at which it can be solved determines the minimal sampling time of the system. In this chapter a novel and computationally efficient search tree algorithm is presented that significantly improves this point-location search for piecewise control laws defined over a large number of (possibly overlapping) polyhedra. Furthermore, the required off-line preprocessing is low and so the approach can be applied to very complex controllers.

11.1 Introduction

In this chapter the *point-location* or *set membership problem* [Sno97] for the class of discrete-time control problems with linear state and input constraints for which an explicit time-invariant piecewise state feedback control law over a set of possibly overlapping polyhedral regions is given, is considered.

The point-location problem comes into play on-line when evaluating the control law. One must identify the state space region in which the measured state lies at the current sampling instance. As the number of defining regions grows, a purely *sequential search* (also known as *exhaustive search*) through the regions is not sufficient to achieve high sampling rates. Hence, it is important to find an efficient on-line search strategy in order to evaluate the control action 'in time' without the need of a heavy additional memory and preprocessing demand.

This work is motivated, but not limited, by the recent developments in the field of controller synthesis for hybrid systems [vS00, Hee99, Son81, BBM00, Bor03, Joh03]. As in most parts of this manuscript, here we consider the class of constrained discrete-time *piecewise affine* (PWA) systems [Son81] that are obtained by partitioning the extended state-input space into polyhedral regions and associating with each region a different affine state update equation, cf. Chapter 4.

As shown in Part II, for PWA systems the *constrained finite time optimal control* (CFTOC) problem can be solved by means of multi-parametric programming and the resulting solution is a time-varying PWA state feedback control law. If the solution to the CFTOC problem is used in a *receding horizon control* [MRRS00, Mac02, GSD05] strategy the time-varying PWA state feedback control law becomes time-invariant and can serve as a control 'lookup table' on-line, thus enabling receding horizon control to be used for fast sampled systems. However, due to the combinatorial nature of the problem the number of state space regions over which the control lookup table is defined grows in the worst case exponentially [Bor03, BMDP02] and therefore efficient on-line search strategies are required to achieve fast sampling rates.

In this chapter a novel, computationally efficient algorithm is presented that performs the aforementioned point-location search for *general* closed-form piecewise (possibly nonlinear) state feedback control laws defined over a finite number of polyhedra or over a finite number of regions for which a *bounding box*[1] [BFT04] computation is feasible. Moreover, control laws that do not form a polyhedral partition, but are composed of a collection of *overlapping* polytopic sets, are included naturally in the algorithm. The proposed point-location search algorithm offers a significant improvement in computation time at the cost of a low additional memory storage demand and very low pre-computation time for the construction of the search tree. This enables the algorithm to work for controller partitions with a large number of regions, which is demonstrated on numerical examples. In order to show its efficiency, the algorithm is compared with the procedure proposed in [TJB03], where a *binary search tree* is pre-computed over the controller state space partition.

11.2 Software Implementation

The presented algorithm is implemented in the Multi-Parametric Toolbox (MPT) [KGB04] for MATLAB®. The toolbox can be downloaded free of charge at: http://control.ee.ethz.ch/~mpt/

---

[1] A bounding box is the (minimum-volume) hyper-rectangle that contains the given set.

11.3 Point Location Problem

We now consider *arbitrary* discrete-time control problems with a closed-form (possibly nonlinear) time-invariant piecewise state feedback control law of the form

$$\mu(x(t)) := \mu_i(x(t)), \qquad \text{if } x(t) \in \mathcal{P}_i, \tag{11.1}$$

where $i = 1, \ldots, N_\mathcal{P}$. $x(t) \in \mathbb{R}^{n_x}$ denotes the state of the controlled system at time $t \geq 0$, $\mu_i(\cdot) \in \mathbb{R}^{n_u}$ are nonlinear control functions (or oracles), and the sets $\mathcal{P}_i$ are compact and possibly *overlapping*, i.e. there exists $\mathcal{P}_i$ and $\mathcal{P}_j$ with $i \neq j$ such that $\mathcal{P}_i \cap \mathcal{P}_j$ is full-dimensional. Moreover, $\mathcal{P} := \{\mathcal{P}_i\}_{i=1}^{N_\mathcal{P}}$ denotes the collection of sets $\mathcal{P}_i$.

In an on-line application the control action $u(t) \in \mathbb{R}^{n_u}$ is implied by

$$u(t) = \mu(x(t)).$$

In order to evaluate the control one needs to identify the state space region $\mathcal{P}_i$ in which the measured state $x(t)$ lies at the sampling instance t, i.e.

Algorithm 3.3 (Control evaluation)

1. measure the state $x(t)$ at time instance t
2. search for the index set of regions $\mathcal{I}$ such that $x(t) \in \mathcal{P}_i$ for all $i \in \mathcal{I}$
   IF $\mathcal{I} = \emptyset$ **THEN** problem infeasible **STOP**
   IF $|\mathcal{I}| > 1$ **THEN** pick one element $i^\star \in \mathcal{I}$
3. apply the control input $u(t) = \mu_{i^\star}(x(t))$ to the system
4. wait for the new sampling time $t + 1$, goto (1.)x

The second step in Algorithm 3.3 is also known as the *point-location* or the *set membership* problem [Sno97]: in other words, given a point $x \in \mathbb{R}^{n_x}$ and a set of sets $\{\mathcal{P}_i\}_{i=1}^{N_\mathcal{P}}$, the goal is to list the set of indices $\mathcal{I}$ such that $x \in \mathcal{P}_i$ for all $i \in \mathcal{I}$. $|\mathcal{I}|$ denotes the cardinality of the discrete set $\mathcal{I}$.

11.4 Constrained Finite Time Optimal Control of Piecewise Affine Systems

A brief overview about interesting problem classes, where point-location plays an important role, is given in the following. The reader is referred to Part II for further detail.

Consider the class of discrete-time, stabilizable, linear hybrid systems that can be described as constrained *piecewise affine* (PWA) systems of the following form

11.4 Constrained Finite Time Optimal Control of Piecewise Affine Systems

$$x(t+1) = f_{\text{PWA}}(x(t), u(t))$$
$$:= A_d x(t) + B_d u(t) + a_d, \quad \text{if } \begin{bmatrix} x(t) \\ u(t) \end{bmatrix} \in \mathcal{D}_d, \quad (4.1)$$

where $t \geq 0$, the domain $\mathcal{D} := \cup_{d=1}^{N_D} \mathcal{D}_d$ of $f_{\text{PWA}}(\cdot, \cdot)$ is a non-empty compact set in $\mathbb{R}^{n_x + n_u}$ with $N_D < \infty$ the number of system dynamics, and $\{\mathcal{D}_d\}_{d=1}^{N_D}$ denotes a polyhedral partition of the domain $\mathcal{D}$, where the closure of $\mathcal{D}_d$ is given by $\bar{\mathcal{D}}_d := \{[\begin{smallmatrix} x \\ u \end{smallmatrix}] \in \mathbb{R}^{n_x + n_u} \mid D_d^x x + D_d^u u \leq D_d^0\}$ and $\text{int}(\mathcal{D}_d) \cap \text{int}(\mathcal{D}_j) = \emptyset$, for all $d \neq j$.

Linear state and input constraints of the general form $C^x x + C^u u \leq C^0$ are naturally incorporated in the description of $\mathcal{D}_d$.

As an example let us define for the aforementioned piecewise affine system (4.1) the *constrained finite time optimal control* (CFTOC) problem

$$J_T^*(x(0)) := \min_{U_T} J_T(x(0), U_T) \quad (6.1a)$$

$$\text{subj. to } \begin{cases} x(t+1) = f_{\text{PWA}}(x(t), u(t)) \\ x(T) \in \mathcal{X}^f, \end{cases} \quad (6.1b)$$

where

$$J_T(x(0), U_T) := \ell_T(x(T)) + \sum_{t=0}^{T-1} \ell(x(t), u(t)) \quad (6.1c)$$

is the cost function (also called performance index), $\ell(\cdot, \cdot)$ the stage cost, $\ell_T(\cdot)$ the final penalty function, $U_T := \{u(t)\}_{t=0}^{T-1}$ is the optimization variable defined as input sequence, $T < \infty$ is the prediction horizon, and $\mathcal{X}^f$ is a compact terminal set in $\mathbb{R}^{n_x}$. The CFTOC problem (6.1a)–(6.1c) implicitly defines the set of feasible initial states $\mathcal{X}_T \subset \mathbb{R}^{n_x}$ ($x(0) \in \mathcal{X}_T$).

Consider the two following restrictions to the aforementioned CFTOC problem (6.1a)–(6.1c)

Problem 6.1 (PWA system, 1-/∞-norm based cost). Let the cost function (6.1c) be composed of

$$\ell(x(t), u(t)) := \|Qx(t)\|_p + \|Ru(t)\|_p \quad \text{and}$$
$$\ell_T(x(T)) := \|Px(T)\|_p, \qquad \square$$

where $\|\cdot\|_p$ with $p \in \{1, \infty\}$ denotes the standard vector 1-/∞-norm [HJ85], and

Problem 10.1 (Constrained LTI system, quadratic cost).

$$f_{\text{PWA}}(x(t), u(t)) := Ax(t) + Bu(t), \quad \text{if } \begin{bmatrix} x(t) \\ u(t) \end{bmatrix} \in \mathcal{D},$$
$$\ell(x(t), u(t)) := x(t)'Qx(t) + u(t)'Ru(t), \quad \text{and}$$
$$\ell_T(x(T)) := x(T)'Px(T). \qquad \square$$

In both, CFTOC Problem 6.1 and Problem 10.1, the solution is a time-varying piecewise affine state feedback control law defined over a polyhedral partition, which is stated in the following theorem, confer also Chapter 6 and [Bor03].

Theorem 6.5 (Solution to CFTOC). The solution to the optimal control problem (6.1a)–(6.1b), restricted to Problem 6.1 or 10.1, is a time-varying piecewise affine function of the initial state $x(0)$

$$\mu(x(0),t) = K_{T-t,i}\,x(0) + L_{T-t,i}, \quad \text{if} \quad x(0) \in \mathcal{P}_i \qquad (6.5)$$

with $u^*(t) = \mu(x(0),t)$, where $t = 0,\ldots,T-1$ and $\{\mathcal{P}_i\}_{i=1}^{N_P}$ is a polyhedral partition of the set of feasible states $x(0)$

$$\mathcal{X}_T = \cup_{i=1}^{N_P} \mathcal{P}_i,$$

with the closure of $\mathcal{P}_i$ given by $\bar{\mathcal{P}}_i = \{x \in \mathbb{R}^{n_x} \mid P_i^x x \leq P_i^0\}$. ∎

In the case that a *receding horizon* (RH) control policy is used in closed-loop, the control is given as a time-invariant state feedback control law of the form

$$\mu_{RH}(x(t)) := \mu(x(t),0) = K_{T,i}\,x(t) + L_{T,i}, \quad \text{if} \quad x(t) \in \mathcal{P}_i, \qquad (6.14)$$

where $i = 1,\ldots,N_P$ and $u(t) = \mu_{RH}(x(t))$ for $t \geq 0$. Note that the closed-form receding horizon control law (6.14) is a special case of control law (11.1), considered in this chapter.

11.5 Alternative Search Approaches

Due to the combinatorial nature of the CFTOC problem (6.1a)–(6.1c), the controller complexity, or the number N_P of state space regions $\mathcal{P}_i$, can grow exponentially with its parameters in the worst case [Bor03, BMDP02]. Hence, for general control problems, a purely sequential search through the regions is not sufficient in an on-line application. It is therefore important to utilize efficient on-line search strategies in order to evaluate the control action 'in time' without the need of a heavy additional memory demand.

Several authors addressed the point-location/memory storage issue but with moderate success for geometrically complex regions or controllers defined over a large number of regions. A few interesting ideas are mentioned in the following. For the solution to the particular CFTOC Problem 6.1 when, additionally, the system is constrained and linear, i.e.

$$f_{PWA}(x(t),u(t)) := Ax(t) + Bu(t), \quad \text{with} \quad \begin{bmatrix} x(t) \\ u(t) \end{bmatrix} \in \mathcal{D}, \qquad (11.2)$$

11.5 ALTERNATIVE SEARCH APPROACHES

the authors in [BBBM01] propose a search algorithm based on the convexity of the piecewise affine value function. Even though this algorithm reduces the storage space significantly, the storage demand as well as the search time are still linear in the number of regions. [JGR05] addresses this issue for the same CFTOC problem class by demonstrating a link between the piecewise affine value function of [BBBM01] and power diagrams (extended Voronoi diagrams). Utilizing standard Voronoi search methods the search time then reduces to $O(\log(N_\mathcal{P}))$.

To the author's knowledge only two other approaches tackle the more general problem, where the only restriction is that the domain of the control law is a non-overlapping polyhedral partition of the state space. (Note that this is more restrictive than the algorithm presented here, confer the introduction of this chapter and Section 11.6.) [GTM04, GTM03] aim at pre-computing a minimal polyhedral representation of the original controller partition in order to reduce storage and search complexity. However, the computation is 'practically' limited to a small number of regions with a small number of facets (see Definition 1.26), since the pre-computation time grows exponentially. Relaxations to a larger number of regions is possible at the cost of data storage and a higher search complexity.

An alternative approach, which will be used here for comparison, was proposed by Tøndel et al. in [TJB03], where a binary search tree is constructed on the basis of the geometric structure of the polyhedral partition[2] by utilizing the facets of the regions as separating hyperplanes to divide the polyhedral partition at each tree level. This however, can lead to a worst case combinatorial number of subdivisions of existing regions and therefore to an additional increase in the number of regions to be considered during the search procedure. The on-line point-location search time is in the best case logarithmic in the number of regions $N_\mathcal{P}$, but worst case linear in the total number of facets, which makes the procedure equivalent to sequential search in the worst case. Moreover, note that the total number of facets, N_F, is typically larger than the original number of regions in the partition, i.e. $N_F > N_\mathcal{P}$. Although the scheme works very well for polyhedral partitions that have a 'simple' geometric structure and/or have a small number of regions, it is computationally prohibitive in the preprocessing time for more complex controller partitions. This is due to the fact that the first step of the pre-processing is to determine on which side of every facet defining hyperplane each region lies, which requires $2N_F N_\mathcal{P}$ linear programs, thereby making this method untenable for moderate to large problems, i.e. greater than 10 000 regions, cf. Section 11.8.3. The memory stor-

---

[2] Even though in the introduction of [TJB03] it is mentioned that overlapping regions and 'holes' in the domain of the controller are handled by the proposed algorithm, these cases are not explicitly treated in the algorithm nor it is directly apparent how this will influence the complexity of the algorithm.

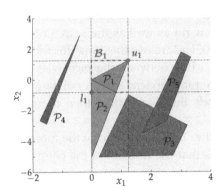
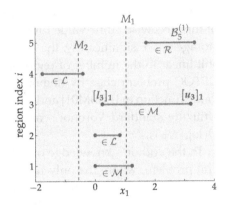

Fig. 11.1: Overlapping collection of polytopic sets $\{\mathcal{P}_i\}_{i=1}^5$ with bounding box $\mathcal{B}_1$ of $\mathcal{P}_1$.

Fig. 11.2: Projection $\mathcal{B}_i^{(1)}$ of the bounding boxes of the polytopic set-collection $\{\mathcal{P}_i\}_{i=1}^5$ of Figure 11.1 onto the x_1-space sorted by the region's index. Indicators for the construction of the first node level of the first dimension of the interval tree are represented in green.

age requirement for the binary search tree is (in the worst case) in the order of $n_X N_F$.

11.6 The Proposed Search Algorithm

The proposed search algorithm is based on minimal volume *bounding boxes* $\mathcal{B}_i$ for each region $\mathcal{P}_i$, which are defined as

$$\mathcal{B}_i := \{x \in \mathbb{R}^{n_x} \mid l_i \leq x \leq u_i\},$$

where the lower and upper bounds l_i and u_i are given by

$$(l_i, u_i) := \arg\min_{l,u} \ \text{vol}\,(\mathcal{B}(l,u))$$

$$\text{subj. to } \mathcal{B}(l,u) = \{x \in \mathbb{R}^{n_x} \mid l \leq x \leq u\} \supseteq \mathcal{P}_i.$$

In other words, $\mathcal{B}_i$ is the 'smallest' axis-aligned n_x-dimensional hyper-rectangle that contains $\mathcal{P}_i$. An example bounding box $\mathcal{B}_1$ can be seen in Figure 11.1.

Remark 11.1. Note that if the regions $\mathcal{P}_i$ are polytopes, then a minimal volume bounding box can be computed using $2n_x$ linear programs of dimension n_x [BFT04]. □

For a given query point, or measured state $x(t)$, the proposed algorithm operates in two stages. First, a list $\mathcal{I}^\mathcal{B}$ of bounding boxes containing the point $x(t)$

is computed, i.e. $x(t) \in \mathcal{B}_i$ for all $i \in \mathcal{I}^\mathcal{B}$ (Section 11.6.1). Second, for each index $i \in \mathcal{I}^\mathcal{B}$, the region $\mathcal{P}_i$ is tested to determine if it contains $x(t)$ (Section 11.6.2). In the following $x(t)$ is simply denoted by x for brevity.

The first stage of this procedure is extremely efficient and computationally 'inexpensive', since the containing bounding boxes can be reported in logarithmic time. This can be done by breaking the search down into one-dimensional range queries, which is possible due to the axis-aligned nature of the bounding boxes. The complexity of the second stage of the algorithm is a function of the overlap between the bounding boxes of adjacent regions. A significant advantage of the proposed search tree is a very simple and effective preprocessing step, which allows the method to be applied to controllers defined over a very large number of regions, i.e. several tens of thousands. As is shown in Section 11.8, there are several large problems of interest to control which have a structure that makes this procedure efficient.

Remark 11.2 (Overlapping regions). Note that overlapping regions are treated naturally and without any additional heuristics by the algorithm. □

11.6.1 Bounding Box Search Tree

In this section, the procedure for reporting the set of indices $\mathcal{I}^\mathcal{B}$ of all bounding boxes that contain a given point x will be detailed. The algorithm relies on the fact that one can decompose the search of a query point $x \in \mathbb{R}^{n_x}$ in a set of bounding boxes in an n_x-dimensional space into n_x separate one-dimensional sequential or parallel searches, because the bounding boxes are all axis-aligned.

The basic steps for constructing the search tree are given in Algorithm 11.3.

Algorithm 11.3 (Building the search tree)

1. compute the bounding box $\mathcal{B}_i$ for each $\mathcal{P}_i$
2. project each bounding box $\mathcal{B}_i$ onto each dimension $d = 1, \ldots, n_x$: define $\mathcal{B}_i^{(d)}$ as the resulting interval
3. build an n_x-dimensional interval tree

Note that step 2 of Algorithm 11.3 for axis-aligned bounding boxes is merely a coordinate extraction of the corner points l_i and u_i.

The proposed search algorithm is an extension to the well known concept of *interval trees* [dSvO00, CLRS01]. Standard interval trees are efficiently ordered binary search trees for determining a set of possibly overlapping one-dimensional line segments that contain a given point or line segment. Consider Figure 11.2, in which the intervals of the bounding boxes in the first dimension for the example in Figure 11.1 are shown. The intervals are spread vertically, ordered by their respective index i, to make them easier to see.

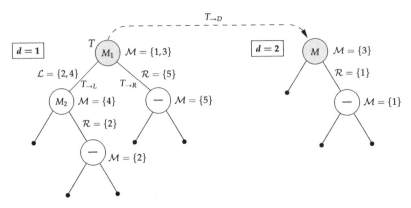

Fig. 11.3: Two-dimensional interval tree for the collection of polytopes $\mathcal{P}$ in Figure 11.1. The gray indicated node in $d = 1$ is explored further in $d = 2$.

Each node of the search tree, cf. Figure 11.3 and 11.2, is associated with a median point M. For example the root node T in Figure 11.3 is associated with the point M_1 in Figure 11.2. The node splits the set of intervals into three sets: The set $\mathcal{L}$, consisting of those entirely on the left of the point M, $\mathcal{R}$ those entirely on the right and $\mathcal{M}$, those that intersect it. The set $\mathcal{M}$ is stored in the node and the left and right branches of the tree are formed by choosing points above and below M and repeating this procedure on $\mathcal{L}$ and $\mathcal{R}$, respectively. By choosing the point M to be the median

$$M := \frac{1}{2}\left(\min_{i \in \mathcal{J}}\{[l_i]_d\} + \max_{i \in \mathcal{J}}\{[u_i]_d\}\right)$$

of the considered intervals $\mathcal{J}$ at a given step, the number of intervals at each level of the tree drops logarithmically. This standard interval tree for the example in Figure 11.2 is shown in the left ($d = 1$) of Figure 11.3. Please note that $[z]_d$ refers to the d-th component of some vector z.

The tree can then be used on-line to determine the set $\mathcal{I}^B$ of intervals containing a given point $[x]_1$, which is the first dimension of the query point x, as follows. Beginning at the root node T, the point $[x]_1$ is compared to the point M_1 associated with the root node. If one assumes that the point $[x]_1$ is larger than M_1, then it is contained in all intervals in the set $\mathcal{M}$ whose right endpoint $[u_i]_1$ is larger than $[x]_1$, since M_1 is less than $[x]_1$ and is also contained in the interval. Note that this search over the set $\mathcal{M}$ can be done in logarithmic time by pre-sorting the endpoints of the intervals in $\mathcal{M}$. Finally, the tree is followed down the right branch, denoted $T_{\to R}$ in Figure 11.3, and this procedure is repeated recursively. If the point $[x]_1$ is less than M_1, then a similar procedure is carried out on the lower bounds and the left branch is followed, which is labeled $T_{\to L}$ in Figure 11.3.

Now this standard method is extended to higher dimensions by building an interval tree over the sets $\mathcal{M}$ at each node using the next dimension $[x]_2$, i.e. $d = 2$. In Figure 11.3, the tree on the left resembles the interval tree for the first measured dimension $[x]_1$. The root node T contains several elements $\mathcal{M} = \{1,3\}$, i.e. $|\mathcal{M}| > 1$, and therefore an interval tree, labeled $T_{\to D}$ in Figure 11.3, over the second dimension ($d = 2$) is constructed for this node, in which only the elements $\{1,3\}$ are considered and where the search is performed for the second dimension of the measured variable $[x]_2$ only. This tree is shown on the right of the figure. By continuing in this manner, the approach is extended to arbitrary dimensions n_x.

11.6.2 Local Search

As already mentioned, the interval search tree only provides a list of candidates $\mathcal{I}^B$ that are possible solutions to the point-location problem. In order to identify the particular index set $\mathcal{I} \subseteq \mathcal{I}^B$ that actually contains the measured point $x(t)$, cf. step 2 of Algorithm 3.3, a local search algorithm needs to be executed on the list of candidate regions by exhaustively testing a set membership $x \in \mathcal{P}_i$ for all $i \in \mathcal{I}^B$.

If the cost function associated with a solution of a CFTOC Problem 6.1 is convex, one can use the approach of [BBBM01] in which the local search can be performed in $(2n_x + 2)|\mathcal{I}^B|$ arithmetic operations.

11.7 Complexity

Preprocessing

The preprocessing phase for the proposed algorithm occurs in two steps. First, the bounding boxes for each region are computed, and then the n_x-dimensional interval tree is built. The calculation of a bounding box requires two linear programs per dimension per region. Therefore, if there are $N_\mathcal{P}$ regions, then the calculation of the bounding boxes requires exactly $2n_x N_\mathcal{P}$ linear programs of dimension n_x. The construction of the interval tree can be performed in $O(n_x N_\mathcal{P} \log(N_\mathcal{P}))$ [dSvO00] and as can be seen from the examples in Section 11.8, the required computation is insignificant compared to the computation of the bounding boxes.

Note that as the preprocessing for this algorithm requires two linear programs per region, it is guaranteed to take significantly less time than the initial computation of the controller. It follows that this approach can be applied to any system for which an explicit controller can be calculated. Note also that bounding boxes are computed in some parametric solvers as the solution is computed [KGB04], making the additional off-line computation negligible.

Storage

The algorithm requires the storage of the defining inequalities for each region as well as the structure of the tree. The tree has a 3-ary structure, two branches for the left and right, and one for the next dimension. In each non-leaf node of the tree is stored a median point M_i and pointers for each of the three branches, totaling four numbers. The leaf nodes then store the indices of the bounding boxes that will be checked during the local search. A completely unbalanced tree in which each leaf node contains exactly one bounding box is the most space consuming configuration. This 'worst case' tree would have $N_P - 1$ non-leaf nodes and N_P leaves for a worst case total storage requirement of $4(N_P - 1) + N_P$ numbers (pointers or reals). Note that this worst case complexity is linear in the number of regions and independent of the state dimension.

On-line Complexity

The interval tree can be traversed in $O(\log(N_P) + |\mathcal{I}^B|)$ time, where $|\mathcal{I}^B|$ is the number of intervals returned [dSvO00]. However, all current methods of doing the secondary search over the list of $|\mathcal{I}^B|$ potential regions returned must be done in linear time. The worst-case complexity is therefore determined by the maximum number of regions that can be returned by the interval tree search, or equally the maximum number of bounding boxes that contain a single point. In the worst case, a point would exist that is contained in every bounding box and therefore the local search for this case would in fact be a complete global search and the resulting worst-case efficiency would be $|\mathcal{I}^B| = N_P$. It is demonstrated by example in the following section that there exist control problems for which the proposed method offers a significant improvement over current approaches.

11.8 Examples

11.8.1 Constrained LTI System

The proposed algorithm of Section 11.6 was applied to the following linear system with three states and two inputs

$$x(t+1) = \begin{bmatrix} 7/10 & -1/10 & 0 \\ 1/5 & -1/2 & 1/10 \\ 0 & 1/10 & 1/10 \end{bmatrix} x(t) + \begin{bmatrix} 1/10 & 0 \\ 1/10 & 1 \\ 1/10 & 0 \end{bmatrix} u(t).$$

11.8 Examples

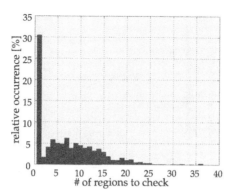

Fig. 11.4: Histogram of the relative occurrence of the number of 'local' regions for the example of Section 11.8.1.

The system is subject to input constraints, $-5\mathbb{1}_2 \leq u(t) \leq 5\mathbb{1}_2$, and state constraints, $-20\mathbb{1}_3 \leq x(t) \leq 20\mathbb{1}_3$. The CFTOC Problem 6.1 is solved with $p = 1$, $T = 8$, $Q = I_3$, $R = \frac{1}{10} I_2$, and $P = 0_{3 \times 3}$. The receding horizon state feedback control law (6.14) consists of 2 568 polyhedral regions in $\mathbb{R}^3$.

As can be seen from Table 11.1, the algorithm presented in Section 11.6 is required to solve $2 \cdot 3 \cdot 2\,568 = 15\,408$ linear programs in the preprocessing phase and needs to store 15 408 real numbers to represent the bounding boxes, as well as 4 424 pointers in order to represent the tree. Since the cost function for this example is piecewise affine and convex, it is possible to use the method in [BBBM01] for the local search, cf. Section 11.6.2, which requires an additional storage of 10 272 real numbers.

In comparison, the binary search tree of [TJB03] for this case consists of 815 unique hyperplanes. For each such hyperplane $2N_\mathcal{P}$ LPs must be solved in the preprocessing phase to compute the index set which corresponds to 4 185 840 linear programs. An actual additional 1 184 782 linear programs are needed to construct the tree, which does not correspond to the worst case scenario.

In order for the proposed method to identify the control law on-line, one has to perform 707 floating point operations to traverse the interval tree in the worst case. Since the tree only provides a necessary condition for the point-location problem, one has to perform a local search on the regions identified by the tree as possible candidates (Section 11.6.2). To provide a worst case bound, an exhaustive check for all possible intersections of the intervals stored in the presented tree was performed. In the worst case 36 regions need to be checked using the method of [BBBM01], which corresponds to 216 flops. However, as can be seen from Figure 11.4, a unique control law is automatically reported by the here proposed search tree in 31 % of all cases without the requirement of doing a secondary local search. In addition, approximately 90 % of all search queries do not require an exhaustive check of more than 15 regions.

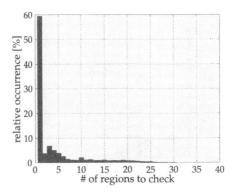

Fig. 11.5: Histogram of the relative occurrence of the number of 'local' regions for the example of Section 11.8.2.

11.8.2 Constrained PWA System

Consider the following piecewise affine system from [MR03]

$$x(t+1) = \begin{cases} A_1 x(t) + Bu(t), & \text{if } [0, 1, 0]x(t) \leq 1, \\ A_2 x(t) + Bu(t) + a, & \text{otherwise,} \end{cases}$$

where

$$A_1 = \begin{bmatrix} 1 & 1/2 & 3/10 \\ 0 & 1 & 1 \\ 0 & 0 & 1 \end{bmatrix}, \quad A_2 = \begin{bmatrix} 1 & 1/5 & 3/10 \\ 0 & 1/2 & 1 \\ 0 & 0 & 1 \end{bmatrix}, \quad B = \begin{bmatrix} 0 \\ 0 \\ 1 \end{bmatrix}, \quad \text{and } a = \begin{bmatrix} 3/10 \\ 1/2 \\ 0 \end{bmatrix}.$$

The system is subject to input constraints, $-1 \leq u(t) \leq 1$, and state constraints, $[-10, -5, -10]' \leq x(t) \leq [10, 5, 10]'$. With $p = 1$, $T = 7$, $Q = I_{3 \times 3}$, $R = 1/10$, and $P = 0_{3 \times 3}$. The solution to the CFTOC Problem 6.1 resulted in a receding horizon state feedback control law (6.14) defined over 2 222 polyhedral regions in $\mathbb{R}^3$.

The off-line construction of the interval search tree for this example required 13 332 LPs to be solved, compared to $7.9 \cdot 10^6$ linear programs which are needed to construct the binary search tree of [TJB03] (this does not correspond to the worst case scenario). Since the cost function of a given CFTOC solution is not

	sequential search	Alg. in [TJB03]	Alg. 11.3 ([BBBM01] locally)
number of LPs (off-line)	—	5 370 622	15 408
runtime (off-line)	—	10 384 secs	10 secs
on-line arithmetic operations (worst case)	106 295	110	923

Tab. 11.1: Computational complexity for the example of Section 11.8.1.

11.8 Examples

	sequential search	Alg. in [TJB03]	Alg. 11.3
number of LPs (off-line)	—	7 913 462	13 332
runtime (off-line)	—	12 810 secs	4.8 secs
on-line arithmetic operations (worst case)	97 984	352	2 602

Tab. 11.2: Computational complexity for the example of Section 11.8.2.

necessarily convex, one cannot use the method of [BBBM01] to perform the local search, and therefore one must perform a sequential search as outlined in Section 11.6.2 on possible candidates. Using the same methodology as in the previous example, it was found that at most 39 regions have to be searched exhaustively. This, however, takes at most 1 720 flops. The worst case number of floating point operations needed to traverse the interval tree is 882. Moreover, almost 60 % of all search queries result in a unique control law during the first phase of the algorithm (Section 11.6.1), cf. Figure 11.5. Therefore, no additional sequential searches are necessary in these cases. Results on the computational complexity are summarized in Table 11.2.

11.8.3 Ball & Plate System

The mechanical 'Ball & Plate' system was introduced in [Bor03, Her96]. The experiment consists of a ball rolling over a gimbal-suspended plate actuated by two independent motors, cf. Figure 11.6. The control objective is to make the ball follow a prescribed trajectory, while minimizing the control effort. The dynamical model for the y-axis of such a device is given by

$$\dot{x}(t) = \begin{bmatrix} 0 & 1 & 0 & 0 \\ 0 & 0 & 700 & 0 \\ 0 & 0 & 0 & 1 \\ 0 & 0 & 0 & -34.69 \end{bmatrix} x(t) + \begin{bmatrix} 0 \\ 0 \\ 0 \\ 3.1119 \end{bmatrix} u(t), \qquad (11.3)$$

where $x := [y, \dot{y}, \alpha, \dot{\alpha}]'$ is the state. $-30 \le y \le 30$ and $-15 \le \dot{y} \le 15$ are the position and velocity of the ball with respect to the y-coordinate, $-0.26 \le \alpha \le 0.26$ and $-1 \le \dot{\alpha} \le 1$ denote the angular position and angular velocity of the plate, respectively. The input voltage to the motor is assumed to be constrained by $-10 \le u \le 10$. In order to take the tracking requirements into account, the state vector is extended with an additional element, which contains the reference signal, hence the augmented state vector is in $\mathbb{R}^5$. The model (11.3) was then discretized with sampling time $T_s = 0.03$ and a closed-form PWA feedback control law (6.14) was derived for the CFTOC Problem 10.1, where the following parameters $T = 10$, $Q = \text{diag}([6, 1/10, 500, 100, 6])$, $R = 1$, and

Fig. 11.6: Ball & Plate laboratory setup. The ball follows a pre-specified trajectory.

$P = Q$ were considered. The controller obtained using the Multi-Parametric Toolbox [KGB04] for MATLAB® is defined over 22 286 regions in $\mathbb{R}^5$.

The computational results for the respective search trees are summarized in Table 11.3. Due to the high number of regions, the binary search tree of [TJB03] was not applicable to this example (denoted by $\star$ in Table 11.3), since it would require the solution of $3.5 \cdot 10^9$ LPs already in the preprocessing stage to determine the index set before building the binary search tree.

In contrast, in the preprocessing stage, the here proposed algorithm has to solve 222 860 LPs to obtain the bounding boxes for all regions. The overall time needed to construct the complete search tree, including the computation of the bounding boxes, was just 80 seconds. 9 324 pointers are needed to represent the tree structure, and 222 860 floating point numbers are needed to describe the bounding boxes.

To estimate the average and the worst case number of arithmetic operations needed to identify the control law on-line, 10 000 random initial conditions over the whole feasible state space were investigated. It can be seen from the histogram distribution depicted in Figure 11.7 that the search tree algorithm identifies at most 500 regions as possible candidates in 86 % of all tested initial

	sequential search	Alg. in [TJB03]	*Alg. 11.3*
number of LPs (off-line)	—	$3.5 \cdot 10^9$	222 860
runtime (off-line)	—	$\star$	80 secs
on-line arithmetic operations (worst case)	1 400 178	$\star$	208 849

Tab. 11.3: Computational complexity for the example of Section 11.8.3. $\star$ denotes that the algorithm in [TJB03] is not computable for this example.

11.8 EXAMPLES

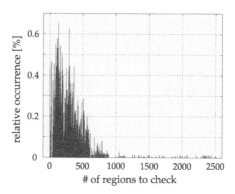

Fig. 11.7: Histogram of the relative occurrence of the number of 'local' regions for the example of Section 11.8.3.

conditions. The subsequent sequential search over 500 regions corresponds to 30 000 floating point operations. In 99% of all tested cases the algorithm identifies at most 1000 regions for subsequent local search, which corresponds to at most 60 000 flops. In the worst case, the search tree will identify as many as 2 544 regions as possible candidates for a sequential search. Notice that this number represents, in the worst case, only 11 % of the total number of regions. This amounts to a maximum of 152 640 flops, whereas traversal of the tree contributes another 56 209 flops. The sequential search through all regions, on the other hand, would require $1.4 \cdot 10^6$ operations and is currently the only other method that can be applied to such a large system. The total number of flops which are needed to be performed on-line is thus (in the worst case) reduced by at least one order of magnitude. To give a sensible feeling for this number of floating point operations, note that a 3 GHz Pentium 4 computer can execute approximately $800 \cdot 10^6$ flops/sec. Given this performance the controlled system can be run at a sampling rate of 4 kHz in the case of the presented search tree, whereas the sequential search has a limit of 500 Hz.

Part IV

Appendix

A

ALTERNATIVE PROOF OF LEMMA 7.7(A)

Note that we do *not need* to impose continuity of $J^*_{\infty,\mathrm{DP}}(x)$ at $x = \mathbb{0}_{n_x}$. It is sufficient to assume that the CITOC solution $J^*_\infty(x)$ is bounded and continuous *at the single point* $x = \mathbb{0}_{n_x}$ and therefore there always exists a K-class function $\bar{J}(\cdot)$ (see Lemma 7.4) bounding $J^*_\infty(\cdot)$ from above, i.e. $J^*_\infty(\cdot) \leq \bar{J}(\|\cdot\|_2)$. We have $0 \leq J_0(\cdot) \leq J^*_\infty(\cdot)$ and according to [Ber01, Sec. 3] $J^*_\infty(\cdot)$ fulfills the Bellman equation. Thus it follows together with the monotonicity property of the operator **T**, cf. [Ber01, Lem 1.1.1], that

$$(\mathbf{T}J_0)(\cdot) \leq (\mathbf{T}J^*_\infty)(\cdot) = J^*_\infty(\cdot).$$

Similarly, $(\mathbf{T}^k J_0)(\cdot) \leq J^*_\infty(\cdot)$ for all $k \geq 0$, and consequently

$$0 \leq J^*_{\infty,\mathrm{DP}}(\cdot) \leq \bar{J}(\|\cdot\|_2).$$

Thus, together with the squeezing theorem of continuity, it is automatic that $J^*_{\infty,\mathrm{DP}}(x)$ is continuous at $x = \mathbb{0}_{n_x}$ and bounded on $\mathcal{X}_\infty$. ∎

BIBLIOGRAPHY

[Art83] ARTSTEIN, Z.: *Stabilization with relaxed controls.* Nonlinear Analysis, 7(11):1163–1173, 1983.

[Bal98] BALAS, E.: *Projection with a minimum system of inequalities.* Computational Optimization and Applications, 10:189–193, 1998.

[Bao05] BAOTIĆ, M.: *Optimal Control of Piecewise Affine Systems – a Multiparametric Approach.* Dr. sc. thesis, ETH Zurich, Zurich, Switzerland, March 2005. Available from http://control.ee.ethz.ch/index.cgi?page=publications;action=details;id=2235.

[BBBM01] BORRELLI, F., M. BAOTIĆ, A. BEMPORAD and M. MORARI: *Efficient On-Line Computation of Constrained Optimal Control.* In *Proc. of the Conf. on Decision & Control*, Orlando, Florida, USA, December 2001.

[BBBM03] BORRELLI, F., M. BAOTIĆ, A. BEMPORAD and M. MORARI: *An Efficient Algorithm for Computing the State Feedback Solution to Optimal Control of Discrete Time Hybrid Systems.* In *Proc. on the American Control Conference*, pages 4717–4722, Denver, Colorado, USA, June 2003.

[BBI01] BURAGO, D., Y. BURAGO and S. IVANOV: *A Course in Metric Geometry*, volume 33 of *Graduate Studies in Mathematics*. American Mathematical Society, Province, Rhode Island, 2001.

[BBM00] BEMPORAD, A., F. BORRELLI and M. MORARI: *Optimal Controllers for Hybrid Systems: Stability and Piecewise Linear Explicit Form*. In *Proc. of the Conf. on Decision & Control*, Sydney, Australia, December 2000.

[BBM02] BEMPORAD, A., F. BORRELLI and M. MORARI: *Model Predictive Control Based on Linear Programming—The Explicit Solution*. IEEE Trans. on Automatic Control, 47(12):1974–1985, December 2002.

[BBM05] BARIĆ, M., M. BAOTIĆ and M. MORARI: *On-line Tuning of Controllers for Systems with Constraints*. In *Proc. of the Conf. on Decision & Control*, pages 8288–8293, Sevilla, Spain, December 2005.

[BCM03] BAOTIĆ, M., F. J. CHRISTOPHERSEN and M. MORARI: *A new Algorithm for Constrained Finite Time Optimal Control of Hybrid Systems with a Linear Performance Index*. In *Proc. of the European Control Conference*, Cambridge, UK, September 2003. Available from http://control.ee.ethz.ch/index.cgi?page=publications.

[BCM06] BAOTIĆ, M., F. J. CHRISTOPHERSEN and M. MORARI: *Constrained Optimal Control of Hybrid Systems With a Linear Performance Index*. IEEE Trans. on Automatic Control, 51(12):1903–1919, December 2006.

[BD62] BELLMAN, R. E. and S. E. DREYFUS: *Applied Dynamic Programming*. Princeton University Press, 2nd printing (1964) edition, 1962.

[BEFB94] BOYD, S., L. EL GHAOUI, E. FERON and V. BALAKRISHNAN: *Linear Matrix Inequalities in System and Control Theory*. Studies in Applied Mathematics. SIAM, 1994.

[Bel57] BELLMAN, R. E.: *Dynamic Programming*. Princeton University Press, Princeton, N.J., 1957.

[Ben03] BENNETT, M.: *Optimal Guidance & Control of Exo-atmospheric Kill Vehicles*. Association for Aeronautics & Astronautics (AAAF), 2003.

[Ber95] BERTSEKAS, D. P.: *Nonlinear Programming*. Athena Scientific, Belmont, Massachusetts, 1995.

[Ber00] BERTSEKAS, D. P.: *Dynamic Programming and Optimal Control*, volume I. Athena Scientific, Belmont, Massachusetts, 2nd edition, 2000.

[Ber01] BERTSEKAS, D. P.: *Dynamic Programming and Optimal Control*, volume II. Athena Scientific, Belmont, Massachusetts, 2nd edition, 2001.

[BF03] BEMPORAD, A. and C. FILIPPI: *Suboptimal explicit RHC via approximate multiparametric quadratic programming*. Journal of Optimization Theory and Applications, 117(1):9–38, April 2003.

[BFT04] BEMPORAD, A., C. FILIPPI and F. D. TORRISI: *Inner and outer approximation of polytopes using boxes*. Computational Geometry: Theory and Applications, 27(2):151–178, 2004.

[BGBM05] BARIĆ, M., P. GRIEDER, M. BAOTIĆ and M. MORARI: *Optimal Control of PWA Systems by Exploiting Problem Structure*. In *IFAC World Congress*, Prague, Czech Republic, July 2005. Available from http://control.ee.ethz.ch/index.cgi?page=publications;action=details;id=2033.

[BGK[+]82] BANK, B., J. GUDDAT, D. KLATTE, B. KUMMER and K. TAMMER: *Non-Linear Parametric Optimization*. Akademie-Verlag, Berlin, 1982.

[BGLM05] BISWAS, P., P. GRIEDER, J. LÖFBERG and M. MORARI: *A Survey on Stability Analysis of Discrete-Time Piecewise Affine Systems*. In *IFAC World Congress*, Prague, Czech Republic, July 2005.

[BGW90] BITMEAD, R. R., M. GEVERS and V. WERTZ: *Adaptive Optimal Control: The Thinking Man's GPC*. International Series in Systems and Control Engineering. Prentice Hall, 1990.

BIBLIOGRAPHY

[Bit88] BITSORIS, G.: *Positively invariant polyhedral sets of discrete-time linear systems.* Int. Journal of Control, 47(6):1713–1726, 1988.

[Bla99] BLANCHINI, F.: *Set invarinace in control — A survey.* Automatica, 35:1747–1767, 1999.

[BM99a] BEMPORAD, A. and M. MORARI: *Control of systems integrating logic, dynamics, and constraints.* Automatica, 35(3):407–427, March 1999.

[BM99b] BEMPORAD, A. and M. MORARI: *Robust Model Predictive Control: A Survey.* In GARULLI, A., A. TESI and A. VICINO (editors): *Robustness in Identification and Control*, volume 245 of *Lecture Notes in Control and Information Sciences*, pages 207–226. Springer-Verlag, 1999.

[BMDP02] BEMPORAD, A., M. MORARI, V. DUA and E. N. PISTIKOPOULOS: *The explicit linear quadratic regulator for constrained systems.* Automatica, 38(1):3–20, January 2002.

[Bor02] BORRELLI, F.: *Discrete Time Constrained Optimal Control.* Dr. sc. techn. thesis, ETH Zurich, Zürich, Switzerland, May 2002.

[Bor03] BORRELLI, F.: *Constrained Optimal Control of Linear and Hybrid Systems*, volume 290 of *Lecture Notes in Control and Information Sciences*. Springer-Verlag, 2003.

[BS79] BAZARAA, M. S. and C. M. SHETTY: *Nonlinear Programming — Theory and Algorithms.* John Wiley & Sons, 1979.

[BS96] BERTSEKAS, D. P. and S. SHREVE: *Stochastic Optimal Control: The Discrete-Time Case.* Athena Scientific, 1996.

[BT00] BLONDEL, V. D. and J. N. TSITSIKLIS: *A survey of computational complexity results in systems and control.* Automatica, 36(9):1249–1274, 2000.

[BT03] BAOTIĆ, M. and F. D. TORRISI: *Polycover.* Technical Report AUT03-11, Automatic Control Laboratory, ETHZ, Switzerland, 2003. Available from http://control.ee.ethz.ch/research/publications/publications.msql?

[BTN01] BEN-TAL, A. and A. NEMIROVSKI: *Lectures on Modern Convex Optimization: Analysis, Algorithms, and Engineering Applications.* MPS/SIAM Series on Optimization. SIAM, 2001.

[BV04] BOYD, S. and L. VANDENBERGHE: *Convex Optimization.* Cambridge University Press, 2004. http://www.stanford.edu/class/ee364/.

[BZ00] BRANICKY, M. S. and G. ZHANG: *Solving Hybrid Control Problems: Level Sets and Behavioral Programming.* In *Proc. on the American Control Conference*, Chicago, Illinois USA, June 2000.

[CB04] CAMACHO, E. F. and C. BORDONS: *Model Predictive Control.* Advanced Textbooks in Control and Signal Processing. Springer-Verlag, London, 2nd edition, 2004.

[Cer63] CERNIKOV, S. N.: *Contraction of finite systems of linear inequalities (in russian).* Doklady Akademiia Nauk SSSR, 152(5):1075–1078, 1963. (English translation in Societ Mathematics Doklady, Vol. 4, No. 5 (1963), pp.1520–1524).

[Chr06] CHRISTOPHERSEN, F. J.: *Optimal Control and Analysis for Constrained Piecewise Affine Systems.* Dr. sc. ETH Zurich thesis, ETH Zurich, Zurich, Switzerland, August 2006. Available from http://e-collection.ethbib.ethz.ch/cgi-bin/show.pl?type=diss&nr=16807.

[CLRS01] CORMEN, T. H., C. E. LEISERSON, R. L. RIVEST and C. STEIN: *Introduction to Algorithms.* MIT Press and McGraw-Hill, 2nd edition, 2001.

[CM07] CHRISTOPHERSEN, F. J. and M. MORARI: *Further Results on 'Infinity Norms as Lyapunov Functions for Linear Systems'*. IEEE Trans. on Automatic Control, 52(3), March 2007.

[CR80] CUTLER, C. R. and B. L. RAMAKER: *Dynamic matrix control - a computer control algorithm*. In *Joint Automatic Control Conference*, volume 1, San Francisco, CA, USA, 1980.

[Dav03] DAVIES, B.: *Exploring Chaos: Theory and Experiment*. Westview Press, 2003.

[DDP99] DAVE, P., F. J. DOYLE III and J. F. PEKNY: *Customization strategies for the solution of linear programming problems arising from large scale model predictive control of a paper machine*. Journal of Process Control, 9(5):385–396, 1999.

[dH94] HERTOG, D. DEN: *Interior Point Approach to Linear, Quadratic and Convex Programming: Algorithms and Complexity*. Number 277 in *Mathematics and its Applications*. Kluwer Academic Publishers, Dordrecht, 1994.

[DP00] DUA, V. and E. N. PISTIKOPOULOS: *An algorithm for the solution of multi-parametric mixed integer linear programming problems*. Annals of Operations Research, 99:123–139, 2000.

[dSvO00] DE BERG, M., O. SCHWARZKOPF, M. VAN KREVELD and M. OVERMARS: *Computational Geometry: Algorithms and Applications*. Springer Verlag, 2nd edition, 2000.

[DV01] DE SCHUTTER, B. and T. VAN DEN BOOM: *On model predictive control for max-min-plus-scaling discrete event systems*. Automatica, 37(7):1049–1056, 2001.

[EM99] EGERSTEDT, M. and C. F. MARTIN: *Trajectory planning in the infinity norm for linear control systems*. Int. Journal of Control, 72(13):1139–1146, September 1999.

[Erd04] ERDEM, G.: *On-line optimization and control of simulated moving bed processes*. Dr. sc. thesis, ETH Zurich, Zurich, Switzerland, December 2004.

[FIAF03] FINDEISEN, R., L. IMSLAND, F. ALLGÖWER and B. A. FOSS: *State and output feedback nonlinear model predictive control: An overview*. European Journal of Control, 9(2–3):190–206, 2003. Survey paper.

[FLL00] FUKUDA, K., T. M. LIEBLING and C. LÜTOLF: *Extended convex hull*. In *Canadian Conference on Computational Geometry*, pages 57–64, July 2000.

[Flo95] FLOUDAS, C. A.: *Nonlinear and Mixed-Integer Optimization*. Oxford University Press, 1995.

[Fre65] FREEMAN, H.: *Discrete-Time Systems: An Introduction to the Theory*. John Wiley & Sons, Inc., 1965.

[FT05] FERRARI-TRECATE, G.: *Hybrid Identification Toolbox (HIT)*, 2005. Available from http://www-rocq.inria.fr/who/Giancarlo.Ferrari-Trecate/HIT_toolbox.html.

[FTCMM02] FERRARI-TRECATE, G., F. A. CUZZOLA, D. MIGNONE and M. MORARI: *Analysis of discrete-time piecewise affine and hybrid systems*. Automatica, 38(12):2139–2146, December 2002.

[FTMLM03] FERRARI-TRECATE, G., M. MUSELLI, D. LIBERATI and M. MORARI: *A clustering technique for the identification of piecewise affine systems*. Automatica, 39(2):205–217, February 2003.

[Fuk97] FUKUDA, K.: *Cdd/cdd+ Reference Manual*, December 1997. Available from www.cs.mcgill.ca/~fukuda/soft/cdd_home/cdd.html.

[Fuk00a] FUKUDA, K.: *Frequently Asked Questions in Polyhedral Computation*, October 2000. Available from http://www.ifor.math.ethz.ch/fukuda/polyfaq/polyfaq.html.

[Fuk00b]   FUKUDA, K.: *Polyhedral computation FAQ*, 2000. Available from http://www.ifor.math.ethz.ch/staff/fukuda.

[Gal80]   GAL, T.: *A "Histogramme" of parametric programming*. Journal of Opl Res. Soc., 31:449–451, 1980.

[Gal95]   GAL, T.: *Postoptimal Analysis, Parametric Programming, and Related Topics*. de Gruyter, Berlin, 2nd edition, 1995.

[GBTM04]   GRIEDER, P., F. BORRELLI, F. D. TORRISI and M. MORARI: *Computation of the constrained infinite time linear quadratic regulator*. Automatica, 40:701–708, 2004.

[Gey05]   GEYER, T.: *Low Complexity Model Predictive Control in Power Electronics and Power Systems*. Dr. sc. thesis, ETH Zurich, Zurich, Switzerland, March 2005. Available from http://control.ee.ethz.ch/index.cgi?page=publications;action=details;id=2124.

[GKBM03]   GRIEDER, P., M. KVASNICA, M. BAOTIĆ and M. MORARI: *Low Complexity Control of Piecewise Affine Systems with Stability Guarantee*. Technical Report AUT03-13, Automatic Control Laboratory, ETHZ, Switzerland, 2003. Available from http://control.ee.ethz.ch/research/publications/publications.msql?.

[GKBM04]   GRIEDER, P., M. KVASNICA, M. BAOTIĆ and M. MORARI: *Low Complexity Control of Piecewise Affine Systems with Stability Guarantee*. In *Proc. on the American Control Conference*, Boston, MA, USA, June 2004.

[GKBM05]   GRIEDER, P., M. KVASNICA, M. BAOTIĆ and M. MORARI: *Stabilizing Low Complexity Feedback Control of Constrained Piecewise Affine Systems*. accepted to Automatica, 2005.

[GLPM03]   GRIEDER, P., M. LÜTHI, P. A. PARRILO and M. MORARI: *Stability & Feasibility of Receding Horizon Control*. In *Proc. of the European Control Conference*, Cambridge, UK, September 2003.

[GMTT04]   GRIMM, G., M. J. MESSINA, S. E. TUNA and A. R. TEEL: *Examples when nonlinear model predictive control is nonrobust*. Automatica, 40(10):1729–1738, October 2004.

[GMTT05]   GRIMM, G., M. J. MESSINA, S. E. TUNA and A. R. TEEL: *Model Predictive Control: For Want of a Local Control Lyapunov Function, All is Not Lost*. IEEE Trans. on Automatic Control, 50(5):546–558, May 2005.

[GPM89]   GARCÍA, C. E., D. M. PRETT and M. MORARI: *Model Predictive Control: theory and practice – A Survey*. Automatica, 25(3):335–348, 1989.

[Gri04]   GRIEDER, P.: *Efficient Computation of Feedback Controllers for Constrained Systems*. PhD thesis, ETH Zurich, Automatic Control Laboratory, 2004.

[Grü00]   GRÜNBAUM, B.: *Convex Polytopes*. Springer-Verlag, 2nd edition, 2000.

[GSD05]   GOODWIN, G. C., M. M. SERON and J. A. DE DONÁ: *Constrained Control and Estimation: An Optimisation Approach*. Communications and Control Engineering. Springer-Verlag, London, 2005.

[GT91]   GILBERT, E. G. and K. T. TAN: *Linear systems with state and control constraints: The theory and application of maximal output admissible sets*. IEEE Trans. on Automatic Control, 36(9):1008–1020, September 1991.

[GTM03]   GEYER, T., F. D. TORRISI and M. MORARI: *Efficient Mode Enumeration of Compositional Hybrid Models*. In *Proc. of the Intern. Workshop on Hybrid Systems: Computation and Control*, volume 2623 of *Lecture Notes in Computer Science*, pages 216–232. Springer-Verlag, 2003.

[GTM04]   GEYER, T., F. D. TORRISI and M. MORARI: *Optimal Complexity Reduction of Piecewise Affine Models Based on Hyperplane Arrangements*. In *Proc. on the American Control Conference*, pages 1190–1195, Boston, Massachusetts, USA, June 2004.

[GW94]   GOLUB, G. H. and J. H. WILKINSON: *Ill-conditioned Eigensystems and the Computation of the Jordan Canonical Form*. In PATEL, R. V., A. J. LAUB and P. M. VAN DOOREN (editors): *Numerical Linear Algebra Techniques for Systems and Control*, pages 589–623. IEEE Press, Lane, NJ, USA, 1994.

[Hah67]   HAHN, W.: *Stability of Motion*, volume 138 of *Die Grundlehren der mathematischen Wissenschaften in Einzeldarstellungen*. Springer-Verlag, Heidelberg, 1967.

[HDB01]   HEEMELS, W. P. M. H., B. DE SCHUTTER and A. BEMPORAD: *Equivalence of hybrid dynamical models*. Automatica, 37(7):1085–1091, 2001.

[Hee99]   HEEMELS, W. P. M. H.: *Linear Complementarity Systems: A Study in Hybrid Dynamics*. PhD thesis, Technische Universiteit Eindhoven, Eindhoven, The Netherlands, November 1999.

[Her96]   HERMANN, O.: *Regelung eines ball and plate systems*. Diploma thesis, ETH Zurich, Zurich, Switzerland, 1996.

[HJ85]   HORN, R. A. and C. R. JOHNSON: *Matrix Analysis*. Cambridge University Press, 1985.

[HW79]   HARDY, G. H. and E. M. WRIGHT: *Some Notations*. In *An Introduction to the Theory of Numbers*, chapter 1.6, pages 7–8. Clarendon Press, Oxford, England, 5th edition, 1979.

[HyS]   *Hybrid System Tools*. Web site: http://wiki.grasp.upenn.edu/~graspdoc/wiki/hst/.

[ILO]   ILOG, INC.: *CPLEX User Manual*. Gentilly Cedex, France. http://www.ilog.fr/products/cplex/.

[JGR05]   JONES, C. N., P. GRIEDER and S. V. RAKOVIĆ: *A Logarithmic Solution to the Point Location Problem for Closed-Form Linear MPC*. In *IFAC World Congress*, Prague, Czech Republic, July 2005.

[JKM04]   JONES, C. N., E. C. KERRIGAN and J. M. MACIEJOWSKI: *Equality Set Projection: A new algorithm for the projection of polytopes in halfspace representation*. Technical Report CUED Technical Report CUED/F-INFENG/TR.463, Department of Engineering, Cambridge University, UK, 2004. Available at http://www-control.eng.cam.ac.uk/~cnj22/.

[JKM05]   JONES, C. N., E. C. KERRIGAN and J. M. MACIEJOWSKI: *Lexicographic Perturbation for Multiparametric Linear Programming with Applications to Control*. Automatica, 2005. Submitted.

[JLT06]   JEON, I.-S., J.-I. LEE and M.-J. TAHK: *Impact-Time-Control Guidance Law for Anti-Ship Missiles*. IEEE Trans. on Control Systems Technology, 14(2):260–266, March 2006.

[Joh03]   JOHANSSON, M.: *Piecewise Linear Control Systems: A Computational Approach*, volume 284 of *Lecture Notes in Control and Information Sciences*. Springer-Verlag, 2003.

[Jon05]   JONES, C. N.: *Polyhedral Tools for Control*. PhD thesis, University of Cambridge, Cambridge, U.K., July 2005.

[JR98]   JOHANSSON, M. and A. RANTZER: *Computation of piece-wise quadratic Lyapunov functions for hybrid systems*. IEEE Trans. on Automatic Control, 43(4):555–559, 1998.

BIBLIOGRAPHY 177

[KA02] KOUTSOUKOS, X. D. and P. J. ANTSAKLIS: *Design of stabilizing switching control laws for discrete- and continuous-time linear systems using piecewise-linear Lyapunov functions.* Int. Journal of Control, 75(12):932–945, 2002.

[Kam01] KAMAU, S. I.: *Different Approaches to Modelling of Hybrid Systems.* Forschungsbericht Forschungsbericht 2001.11, Arbeitsbereich Regelungstechnik, TU Hamburg-Harburg, Hamburg, Germany, April 2001.

[Kar84] KARMARKAR, N.: *A new polynomial-time algorithm for linear programming.* Combinatorica, 4:373–395, 1984.

[KAS92] KIENDL, H., J. ADAMY and P. STELZNER: *Vector norms as Lyapunov functions for linear systems.* IEEE Trans. on Automatic Control, 37(6):839–842, June 1992.

[KC01] KOUVARITAKIS, B. and M. CANNON (editors): *Non-linear Predictive Control: Theory and Practice.* The Institution of Engineering and Technology, London, UK, 2001.

[KCHF07] KVASNICA, M., F. J. CHRISTOPHERSEN, M. HERCEG and M. FIKAR: *Polynomial Approximation of Closed-form MPC for Piecewise Affine Systems.* 2007. Submitted.

[Kel02] KELLETT, C. M.: *Advances in Converse and Control Lyapunov Functions.* PhD thesis, University of California, Santa Barbara, CA, USA, 2002.

[Ker00] KERRIGAN, E. C.: *Robust Constraint Satisfaction: Invariant Sets and Predictive Control.* PhD thesis, University of Cambridge, Cambridge, UK, November 2000.

[KGB04] KVASNICA, M., P. GRIEDER and M. BAOTIĆ: *Multi-Parametric Toolbox (MPT)*, 2004. Available from http://control.ee.ethz.ch/~mpt/.

[KGBC06] KVASNICA, M., P. GRIEDER, M. BAOTIĆ and F. J. CHRISTOPHERSEN: *Multi-Parametric Toolbox (MPT).* Automatic Control Laboratory, ETH Zurich, March 2006. Available from http://control.ee.ethz.ch/~mpt/downloads/MPTmanual.pdf.

[KGBM03] KVASNICA, M., P. GRIEDER, M. BAOTIĆ and M. MORARI: *Multi-Parametric Toolbox (MPT).* In *Proc. of the Intern. Workshop on Hybrid Systems: Computation and Control*, volume 2993 of *Lecture Notes in Computer Science*, pages 448–462, Pennsylvania, Philadelphia, USA, March 2003. Springer-Verlag. Available from http://control.ee.ethz.ch/index.cgi?page=publications&action=details&id=53.

[Kha79] KHACHIYAN, L.G.: *A polynomial algorithm in linear programming.* Soviet Mathematics Doklady, 20:191–194, 1979.

[Kha96] KHALIL, H. K.: *Nonlinear Systems.* Prentice Hall, 2nd edition, 1996.

[KKK95] KRSTIĆ, M., I. KANELLAKOPOULOS and P. KOKOTOVIĆ: *Nonlinear and Adaptive Control Design.* Adaptive and Learning Systems for Signal Processing, Communications, and Control. John Wiley & Sons, Inc., 1995.

[KM02] KERRIGAN, E. C. and D. Q. MAYNE: *Optimal control of constrained, piecewise affine systems with bounded disturbances.* In *Proc. of the Conf. on Decision & Control*, pages 1552–1557, Las Vegas, Nevada, USA, December 2002.

[KS90] KEERTHI, S. S. and K. SRIDHARAN: *Solution of parametrized linear inequalities by fourier elimination and its applications.* Journal of Optimization Theory and Applications, 65(1):161–169, 1990.

[KT02] KELLETT, C. M. and A. R. TEEL: *On Robustness of Stability and Lyapunov Functions for Discontinuous Difference Equations.* In *Proc. of the Conf. on Decision & Control*, pages 4282–4287, 2002.

[KT03] KELLETT, C. M. and A. R. TEEL: *Results on Discrete-Time Control-Lyapunov Functions*. In *Proc. of the Conf. on Decision & Control*, pages 5961–5966, Maui, Hawaii, USA, December 2003.

[Laz06] LAZAR, M.: *Model Predictive Control of Hybrid Systems: Stability and Robustness*. PhD thesis, Technical University of Eindhoven, Eindhoven, The Netherlands, September 2006.

[Lev96] LEVINE, W. S. (editor): *The Control Handbook*. The electrical engineering handbook series. CRC press, 1996.

[LHW+05] LAZAR, M., W. P. M. H. HEEMELS, S. WEILAND, A. BEMPORAD and O. PASTRAVANU: *Infinity Norms as Lyapunov functions for Model Predictive Control of Constrained PWA Systems*. In *Proc. of the Intern. Workshop on Hybrid Systems: Computation and Control*, volume 3414 of *Lecture Notes in Computer Science*, Zurich, Switzerland, March 2005. Springer-Verlag.

[LHWB06] LAZAR, M., W. P. M. H. HEEMELS, S. WEILAND and A. BEMPORAD: *Stabilizing Model Predictive Control of Hybrid Systems*. IEEE Trans. on Automatic Control, 51(11):1813–1818, 2006.

[Lin03] LINCOLN, B.: *Dynamic Programming and Time-Varying Delay Systems*. PhD thesis, Department of Automatic Control, Lund Institute of Technology, Sweden, Lund, Sweden, May 2003.

[LL06] LAUMANNS, M. and E. LEFEBER: *Robust optimal control of material flows in demand-driven supply networks*. Physica A: Statistical Mechanics and its Applications, 363(1):24–31, April 2006.

[Löf03] LÖFBERG, J.: *Minimax approaches to robust model predictive control*. PhD thesis, Linköping University, Sweden, April 2003.

[Löf04] LÖFBERG, J.: *YALMIP : A Toolbox for Modeling and Optimization in MATLAB*. In *Proc. of the CACSD Conference*, Taipei, Taiwan, 2004. Available from http://control.ee.ethz.ch/~joloef/yalmip.php.

[LS95] LEWIS, F. L. and V. L. SYRMOS: *Optimal Control*. John Wiley & Sons, 2nd edition, 1995.

[LT85] LANCASTER, P. and M. TISMENETSKY: *The Theory of Matrices*. Computer Science and Applied Mathematics. Academic Press Inc., 2nd edition, 1985.

[LTS99] LYGEROS, J., C. TOMLIN and S. SASTRY: *Controllers for reachability specifications for hybrid systems*. Automatica, 35(3):349–370, 1999.

[Lya92] LYAPUNOV, A. M.: *The general problem of the stability of motion*. Int. Journal of Control, 55:531–773, 1992. This is the English translation of the Russian original published in 1892.

[LZ05] LÖHNE, A. and C. ZĂLINESCU: *On Convergence of Closed Convex Sets*. Technical Report 05-02, Department of Mathematics and Computer Science, Martin-Luther University Halle, Halle, germany, 2005. Available from http://www.mathematik.uni-halle.de/reports/shadows/05-02report.html.

[Mac02] MACIEJOWSKI, J. M.: *Predictive Control with Constraints*. Prentice Hall, 2002.

[May01a] MAYNE, D. Q.: *Constrained Optimal Control*. European Control Conference, Plenary Lecture, September 2001.

[May01b] MAYNE, D. Q.: *Control of Constrained Dynamic Systems*. European Journal of Control, 7:87–99, 2001.

[Md05] MUÑOZ DE LA PEÑA SEQUEDO, D.: *Model Predictive Control for Uncertain Systems*. PhD thesis, Departamento de Ingeniería de Sistemas y Automática, Universidad de Sevilla, Sevilla, Spain, October 2005.

[ML99]   MORARI, M. and J. H. LEE: *Model predictive control: past, present and future.* Computers & Chemical Engineering, 23(4–5):667–682, 1999.

[MP86]   MOLCHANOV, A. P. and E. S. PYATINTSKII: *Lyapunov functions that specify necessary and sufficient conditions of absolute stability of nonlinear nonstationary control systems. III.* Automat. Remote Contr., 47:620–630, 1986. Translated from Avtomatika i Telemekhanika, No. 5, pp. 38–49, May, 1986.

[MR03]   MAYNE, D. Q. and S. V. RAKOVIĆ: *Model predictive control of constrained piecewise affine discrete-time systems.* Int. Journal on Robust and Nonlinear Control, 13(3-4):261–279, 2003.

[MRRS00] MAYNE, D. Q., J. B. RAWLINGS, C. V. RAO and P. O. M. SCOKAERT: *Constrained model predictive control: Stability and optimality.* Automatica, 36(6):789–814, June 2000.

[MRVK06] MAYNE, D. Q., S. V. RAKOVIĆ, R. B. VINTER and E. C. KERRIGAN: *Characterization of the solution to a constrained H_∞ optimal control problem.* Automatica, 42(3):371–382, mar 2006.

[MS72]   MITRA, D. and H. S. SO: *Existence Conditions for L_1 Liapunov Functions for a Class of Nonautonomous Systems.* IEEE Trans. on Circuit Theory, CT-19(6):594–598, November 1972.

[Neu04] NEUMAIER, A.: *Complete Search in Continuous Global Optimization and Constraint Satisfaction.* In ISERLES, A. (editor): *Acta Numerica*, Lecture Notes in Control and Information Sciences. Cambridge University Press, 2004.

[Num02] NUMERICAL ALGORITHMS GROUP, LTD.: *NAG Foundation Toolbox for MATLAB 6.* Oxford, UK, 2002. http://www.nag.co.uk/.

[OH55]   ORCHARD-HAYS, W.: *Notes on linear programming (Part 6): the Rand code for the simplex method (SX4).* Technical Report 1440, Rand Corporation, 1955.

[Par04] PARRILO, P. A.: *Sums of squares of polynomials and their applications.* In *ISSAC '04: Proceedings of the 2004 international symposium on Symbolic and algebraic computation*, pages 1–1, New York, NY, USA, 2004. ACM Press.

[PEN]   PENOPT GBR.: *PENBMI.* http://www.penopt.com/.

[PL03]   PARRILO, P. A. and S. LALL: *Semidefinite programming relaxations and algebraic optimization in Control.* European Journal of Control, 9(2-3), 2003.

[PLYG03] PEREA-LÓPEZ, E., B. E. YDSTIE and I. E. GROSSMANN: *A model predictive control strategy for supply chain optimization.* Computers & Chemical Engineering, 27:1201–1218, 2003.

[Pol95] POLAŃSKI, A.: *On infinity norms as Lyapunov functions for linear systems.* IEEE Trans. on Automatic Control, 40(7):1270–1274, July 1995.

[Pol97] POLAŃSKI, A.: *Lyapunov Function Construction by Linear Programming.* IEEE Trans. on Automatic Control, 42(7):1013–1016, July 1997.

[PPSP04] PRAJNA, S., A. PAPACHRISTODOULOU, P. SEILER and P. A. PARRILO: *SOSTOOLS: Sum of squares optimization toolbox for MATLAB*, 2004.

[Pro63] PROPOI, A. I.: *Use of linear programming methods for synthesizing sampled-data automatic systems.* Automation and Remote Control, 24(7):837–844, 1963.

[QB97]   QIN, S. and T. BADGEWELL. *An overview of industrial model predictive control technology.* Chemical Process Control, 93(316):232–256, 1997.

[Raw00] RAWLINGS, J. B.: *Tutorial Overview of Model Predictive Control.* IEEE Control Systems Magazine, 20(3):38–52, June 2000.

[RBL04] ROLL, J., A. BEMPORAD and L. LJUNG: *Identification of piecewise affine systems via mixed-integer programming.* Automatica, 40:37–50, 2004.

[RH05]　ROZGONYI, S. and K. M. HANGOS: *Hybrid Modelling and Control of an Industrial Vaporizer*. In *Proc. on the Intern. Conf. on Process Control*, Bratislava, Slovakia, June 2005.

[RKM03]　RAKOVIĆ, S. V., E. C. KERRIGAN and D. Q. MAYNE: *Reachability computations for constrained discrete-time systems with state- and input-dependent disturbances*. In *Proc. of the Conf. on Decision & Control*, pages 3905–3910, Maui, Hawaii, USA, December 2003.

[RM05]　RAKOVIĆ, S. V. and D. Q. MAYNE: *A Simple Tube Controller for Efficient Robust Model Predictive Control of Constrained Linear Discrete Time Systems Subject to Bounded Disturbances*. In *Proceedings of the 16th IFAC World Congress IFAC 2005*, Praha, Czech Republic, July 2005. Invited Session.

[Roc97]　ROCKAFELLAR, R. T.: *Convex Analysis*. Princeton University Press, 1997.

[Ros03]　ROSSITER, J. A.: *Model-Based Predictive Control: A Practical Approach*. CRC Press Inc., 2003.

[RRTP76]　RICHALET, J., A. RAULT, J. L. TESTUD and J. PAPON: *Algorithmic control of industrial processes*. In *Proceedings of the Fourth IFAC symposium on identification and system parameter estimation*, pages 1119–1167, 1976.

[RTV97]　ROOS, C., T. TERLAKY and J. P. VIAL: *Theory and Algorithms for Linear Optimization: An Interior Point Approach*. John Wiley & Sons, Inc., New York, 1997.

[Sch86]　SCHRIJVER, A.: *Theory of Linear and Integer Programming*. John Wiley & Sons, Inc., New York, 1986.

[SD87]　SZNAIER, M. and M. J. DAMBORG: *Suboptimal control of linear systems with state and control inequality constraints*. In *Proc. of the Conf. on Decision & Control*, volume 1, pages 761–762, December 1987.

[SDRW01]　SAFFER II, D. R., F. J. DOYLE III, A. RIGOPOULOS and P. WISNEWSKI: *MPC study for a dual headbox CD control problem*. Pulp and Paper Canada, 102(12):97–101, 2001.

[SG54]　SAATY, T. L. and S. I. GASS: *The parametric objective function 1*. Operations research, 2:316–319, 1954.

[SL89]　STOKEY, N. L. and R. E. LUCAS: *Recursive Methods in Economic Dynamics*. Harvard University Press, Cambridge, Massachusetts, 2001 edition, 1989.

[Sno97]　SNOEYINK, J.: *Point Location*. In GOODMAN, J. E. and J. O'ROUKE (editors): *Handbook of Discrete and Computational Geometry*, chapter 30, pages 558–574. CRC Press, Boca Raton, New York, 1997.

[Son81]　SONTAG, E. D.: *Nonlinear regulation: The piecewise linear approach*. IEEE Trans. on Automatic Control, 26(2):346–358, April 1981.

[Son83]　SONTAG, E. D.: *A Lyapunov-Like Characterization of Asymptotic Controllability*. SIAM Journal of Control and Optimization, 21(3):462–471, May 1983.

[STJ05]　SPJØTVOLD, J., P. TØNDEL and T. A. JOHANSEN: *A Method for Obtaining Continuous Solutions to Multiparametric Linear Programs*. In *IFAC World Congress*, Prague, Czech Republic, 2005.

[SX97]　SHAMMA, J. S. and D. XIONG: *Linear nonquadratic optimal control*. IEEE Trans. on Automatic Control, 42:875–879, June 1997.

[TB04]　TORRISI, F. D. and A. BEMPORAD: *HYSDEL – A Tool for Generating Computational Hybrid Models for Analysis and Synthesis Problems*. IEEE TCST – Special Issue on Computer Automated Multi-Paradigm Modeling, 12(2):235–249, March 2004.

[TBB+02]   TORRISI, F. D., A. BEMPORAD, G. BERTINI, P. HERTACH, D. JOST and D. MIGNONE: *HYSDEL 2.0.5 – User Manual*. Technical Report AUT02-10, Automatic Control Laboratory, ETH Zurich, August 2002.

[TJB03]   TØNDEL, P., T. A. JOHANSEN and A. BEMPORAD: *Evaluation of Piecewise Affine Control via Binary Search Tree*. Automatica, 39(5):945–950, May 2003.

[Tøn00]   TØNDEL, P.: *Constrained Optimal Control via Multiparametric Quadratic Programming*. PhD thesis, Department of Engineering Cybernetics, NTNU, Trondheim, Norway, 2000.

[Tor02]   TORRISI, F. D.: *Hybrid System DEscription Language (HYSDEL)*, 2002. Available from http://control.ee.ethz.ch/~hybrid/hysdel/.

[Tor03]   TORRISI, F. D.: *Modeling and Reach-Set Computation for Analysis and Optimal Control of Discrete Hybrid Automata*. Dr. sc. Thesis, ETH Zurich, Zürich, Switzerland, March 2003.

[TW01]   TANG, W. S. and J. WANG: *A Recurrent Neural Network for Minimum Infinity-Norm Kinematic Control of Redundant Manipulators with an Improved Problem Formulation and Reduced Architecture Complexity*. IEEE Trans. on Systems, Man and Cybernetics, Part B, 31(1):98–105, February 2001.

[Vid93]   VIDYASAGAR, M.: *Nonlinear Systems Analysis*. Prentice Hall, 2nd edition, 1993.

[vS00]   VAN DER SCHAFT, A. and H. SCHUMACHER: *An Introduction to Hybrid Dynamical Systems*, volume 251 of *Lecture Notes in Control and Information Sciences*. Springer-Verlag, 2000.

[Wik]   WIKIPEDIA: *Tent map*. http://en.wikipedia.org/wiki/Tent_map.

[ZDG96]   ZHOU, K., J. C. DOYLE and K. GLOVER: *Robust and Optimal Control*. Prentice Hall, 1996.

[Zie95]   ZIEGLER, G. M.: *Lectures on Polytopes*. Graduate Texts in Mathematics. Springer-Verlag, 1995.

[ZW62]   ZADEH, L. A. and L. H. WHALEN: *On optimal control and linear programming*. IEEE Trans. on Automatic Control, AC-7:45–46, January 1962.

AUTHOR INDEX

Adamy, J. IX, 95–98, 105
Allgöwer, F. 136
Antsaklis, P. J. 96
Artstein, Z. 24

Badgewell, T. 26
Balakrishnan, V. 15, 16
Balas, E. 11
Bank, B. 17
Baotić, M. 9–11, 17–19, 32, 33, 41, 46–48, 51, 53, 55, 56, 61, 62, 82, 87, 89, 90, 96, 97, 99, 115, 121, 148, 149, 151, 155, 159, 160–164
Barić, M. 18, 56
Bazaraa, M. S. 3, 15
Bellman, R. E. 53
Bemporad, A. VII, 26, 28, 32, 33, 36, 39, 42, 44, 46, 48, 51, 52, 60, 86, 96, 99, 110, 111, 115, 130, 136, 138, 146, 147, 151, 154–156, 159, 161–164
Ben-Tal, A. 15
Bennett, M. 35
Bertini, G. 42

Bertsekas, D. P. 3, 15, 49, 53, 71–73, 75
Biswas, P. 138
Bitmead, R. R. 46
Bitsoris, G. 96, 105, 110
Blanchini, F. 117
Blondel, V. D. 65
Bordons, C. 26
Borrelli, F. VII, IX, 17–19, 28, 32, 33, 44, 46, 48, 50, 51, 53, 96, 99, 110, 111, 138, 146, 151, 154, 155, 159, 161–163
Boyd, S. 10, 15, 16
Branicky, M. S. 44
Burago, D. 3, 69
Burago, Y. 3, 69

Camacho, E. F. 26
Cernikov, S. N. 11
Christophersen, F. J. XI, 10, 11, 41, 46, 47, 53, 62, 85, 87, 96, 97, 99, 147
Cormen, T. H. 157
Cutler, C. R. 26
Cuzzola, F. A. 123

Damborg, M. J. 46
Dave, P. 34
Davies, B. 118
de Berg, M. 157, 159, 160
De Doná, J. A. 21, 23, 26, 36, 37, 68, 70, 151
De Schutter, B. VII, 39, 44, 51
den Hertog, D. 15
Doyle III, F. J. 34
Doyle, J. C. 33
Dreyfus, S. E. 53
Dua, V. VII, IX, 28, 46, 52, 146, 151, 154

Egerstedt, M. 35, 96
El Ghaoui, L. 15, 16
Erdem, G. 35, 96

Feron, E. 15, 16
Ferrari-Trecate, G. 39, 42, 124
Fikar, M. 147
Filippi, C. 32, 33, 146, 151, 156
Findeisen, R. 136
Floudas, C. A. 15, 17
Foss, B. A. 136
Freeman, H. 21, 23
Fukuda, K. 8, 10–12

Gal, T. 19, 51
García, C. E. 26
Gass, S. I. 19
Gevers, M. 46
Geyer, T. 32, 33, 146, 148, 155
Gilbert, E. G. 21
Glover, K. 33
Golub, G. H. 107
Goodwin, G. C. 21, 23, 26, 36, 37, 68, 70, 151
Grieder, P. 9–11, 32, 33, 41, 46, 47, 55, 56, 61, 62, 82, 87, 89, 90, 97, 115, 118, 119, 127, 138, 146, 148, 149, 151, 155, 160, 164
Grimm, G. 14, 24, 38
Grossmann, I. E. 34
Grünbaum, B. 7, 10
Guddat, J. 17

Hahn, W. 3, 14, 21, 110
Hangos, K. M. 35, 96
Hardy, G. H. XXV
Heemels, W. P. M. H. VII, 36, 39, 44, 49, 51, 86, 96, 110, 138, 151
Herceg, M. 147

Hermann, O. 163
Hertach, P. 42
Horn, R. A. 34, 49, 84, 97–100, 137, 153

ILOG, Inc. 52, 149
Imsland, L. 136
Ivanov, S. 3, 69

Jeon, I.-S. 35
Johansen, T. A. 19, 146, 151, 155, 161–164
Johansson, M. 44, 123, 151
Johnson, C. R. 34, 49, 84, 97–100, 137, 153
Jones, C. N. 11, 17, 19, 51, 58, 155
Jost, D. 42

Kamau, S. I. 39
Kanellakopoulos, I. 24
Karmarkar, N. 15
Keerthi, S. S. 11
Kellett, C. M. 24, 25
Kerrigan, E. C. 11, 19, 21, 46, 51, 58, 136
Khachiyan, L.G. 15
Khalil, H. K. 3, 14, 21
Kiendl, H. IX, 95–98, 105
Klatte, D. 17
Kokotović, P. 24
Koutsoukos, X. D. 96
Krstić, M. 24
Kummer, B. 17
Kvasnica, M. 10, 11, 41, 46, 47, 55, 61, 62, 82, 87, 89, 90, 97, 115, 147–149, 151, 160, 164

Lall, S. 147
Lancaster, P. 98–100
Laumanns, M. 34, 96
Lazar, M. 23, 36, 68, 70, 79, 86, 96, 110, 138
Lee, J. H. 26
Lee, J.-I. 35
Lefeber, E. 34, 96
Leiserson, C. E. 158
Lewis, F. L. 33
Liberati, D. 39
Liebling, T. M. 11
Lincoln, B. 32, 33
Ljung, L. 39
Löfberg, J. 35, 36, 46, 113, 117, 138
Löhne, A. 69

Lucas, R. E. 72
Lüthi, M. 118, 119, 127
Lütolf, C. 11
Lyapunov, A. M. 21
Lygeros, J. 44

Maciejowski, J. M. IX, 11, 19, 26, 51, 58, 110, 151
Martin, C. F. 35, 96
Mayne, D. Q. IX, 11, 35, 36, 46, 50, 53, 110, 115, 117, 131, 133, 136, 151, 162
Messina, M. J. 14, 24, 38
Mignone, D. 42, 123
Mitra, D. 96
Molchanov, A. P. 96
Morari, M. 11, 18, 26, 28, 39, 41, 44, 46, 48, 51, 52, 56, 60, 61, 85–87, 89, 90, 96, 97, 99, 110, 111, 115, 118, 119, 123, 127, 130, 136, 138, 146–149, 151, 154, 155, 159, 161–163
Muñoz de la Peña Sequedo, D. 32, 33
Muselli, M. 39

Nemirovski, A. 15
Neumaier, A. 16
Numerical Algorithms Group, Ltd. 61, 87, 89, 90, 112, 148

Orchard-Hays, W. 19
Overmars, M. 157, 159, 160

Papachristodoulou, A. 147
Papon, J. 26
Parrilo, P. A. 118, 119, 127, 147
Pastravanu, O. 86, 96, 138
Pekny, J. F. 34
PENOPT GbR. 113
Perea-López, E. 34
Pistikopoulos, E. N. VIII, IX, 28, 46, 52, 146, 151, 154
Polański, A. IX, 85, 95–98, 102, 103, 108, 109
Prajna, S. 147
Prett, D. M. 26
Propoi, A. I. 26
Pyatintskii, E. S. 96

Qin, S. 26

Raković, S. V. 11, 131, 133, 136, 162
Ramaker, B. L. 26
Rantzer, A. 123

Rao, C. V. IX, 26, 35, 36, 46, 110, 115, 117, 136, 151
Rault, A. 26
Rawlings, J. B. IX, 26, 35, 36, 46, 110, 115, 117, 136, 151
Richalet, J. 26
Rigopoulos, A. 34
Rivest, R. L. 157
Rockafellar, R. T. 3, 12
Roll, J. 39
Roos, C. 15
Rossiter, J. A. 26
Rozgonyi, S. 35, 96

Saaty, T. L. 19
Saffer II, D. R. 34
Sastry, S. 44
Schrijver, A. 15–17
Schumacher, H. VII, 39, 44, 151
Schwarzkopf, O. 157, 159, 160
Scokaert, P. O. M. IX, 26, 35, 36, 46, 110, 115, 117, 136, 151
Seiler, P. 147
Seron, M. M. 21, 23, 26, 36, 37, 68, 70, 151
Shamma, J. S. 96
Shetty, C. M. 3, 15
Shreve, S. 53
Snoeyink, J. 150, 152
So, H. S. 96
Sontag, E. D. VII, 24, 39, 44, 151
Spjøtvold, J. 19
Sridharan, K. 11
Stein, C. 157
Stelzner, P. IX, 95–98, 105
Stokey, N. L. 72
Syrmos, V. L. 33
Sznaier, M. 46

Tahk, M.-J. 35
Tammer, K. 17
Tan, K. T. 21
Tang, W. S. 35, 96
Teel, A. R. 14, 24, 25, 38
Terlaky, T. 15
Testud, J. L. 26
Tismenetsky, M. 98–100
Tomlin, C. 44
Tøndel, P. 17–19, 32, 33, 146, 151, 155, 161–164
Torrisi, F. D. 11, 33, 42, 46, 146, 148, 151, 155, 156

Tsitsiklis, J. N. 65
Tuna, S. E. 14, 24, 38

Van den Boom, T. 39
van der Schaft, A. VII, 39, 44, 151
van Kreveld, M. 157, 159, 160
Vandenberghe, L. 10, 15
Vial, J. P. 15
Vidyasagar, M. 3, 21, 23, 70, 79, 110
Vinter, R. B. 136

Wang, J. 35, 96
Weiland, S. 36, 86, 96, 110, 138
Wertz, V. 46
Whalen, L. H. 34

Wikipedia 118
Wilkinson, J. H. 107
Wisnewski, P. 34
Wright, E. M. XXV

Xiong, D. 96

Ydstie, B. E. 34

Zadeh, L. A. 34
Zhang, G. 44
Zhou, K. 33
Ziegler, G. M. 7, 10, 12
Zălinescu, C. 69

INDEX

attractivity, **22**, 23, 37
autonomous system, XVIII, **21**, 21–24

ε-ball, *see* set | ε-ball
Bellman
    Bellman equation, 46, 74, 75, 175
    Bellman's optimality principle, 55
binary search tree, 157, 161, 162, 164, 168–171
boundary, *see* set | boundary
bounding box, IV, 162–164, 167, 171

CFTOC, II–IV, 35, 45–47
Chebyshev ball, **10**
    Chebyshev center, 10, 11
    Chebyshev radius, **10**
CITOC, III, 46
closed-form solution, II–IV, 28–32, 38
closure, *see* set | closure
convergence rate, 79, 81
cost
    cost function, II, III, XVIII, 14, 17, 18, **33**, 33–36, 45, 46
    cost-to-go function, 56, 57
    final penalty cost, XVIII, **33**, 46
    linear cost function, 17, **34**,
    quadratic cost function, **33**, 33, 34, 36
    stage cost function, XVIII, **33**,
    terminal cost, *see* cost | final penalty cost

dimension, *see* set | dimension
distance
    Hausdorff distance, 72
DP, *see* dynamic program
    point-to-set distance, 23
dynamic program, III, XVII, XVIII, 46, 56–60, 63, 64, **72**, 72–78, 81, 83, 89–91, 94, 118, 120, 122

equilibrium point, 22–24, 36–38, 40, 41, 46
exhaustive search, *see* squential search
explicit solution, *see* closed-form solution

feasibility, I, III, IV, **18**, 37, 41
    for all time, IV, 27, 31, 35, **36**, 36, 38, 46, 62, 66, 99–101, 114, 115, 118, 119

function
    K-class function, **13**, 14, 24, 37, 71, 73, 83, 144, 175
    KL-class function, **14**
    K_∞-class function, **14**
    L-class function, **14**
    $\widetilde{K}$-class function, **14**
    affine function, 12
    concave function, 13
    convex function, 13
    indefinite function, 13
    Lyapunov function, *see* sability
    negative definite function, 13
    piecewise affine function, **12**, 12, 52, 53, 118, 143, 160
    piecewise quadratic function, 12
    positive definite function, **13**, 23, 24, 33
    value function, XVIII, **17**, 18, 36, 38, 51–53, 56, 61, 63, 66, 68, 69, 72–74, 76, 77, 81–85, 88, 89, 91, 92, 96, 127, 128

half-space, **8**, 8, 9, 12
HIT, 42
hull
    affine hull, XV, **4**, 5
    convex hull, XV, **3**, 9, 12, 107
Hybrid Identification Toolbox, *see* HIT
hybrid system, I, 35, 36, 39, 41, 42, 45
HYbrid System DEscription Language, *see* HYSDEL
hyperplane, 7
HYSDEL, 42

interior, *see* set | interior
    relative interior, *see* set | interior
interval tree, IV, 164–169

Jordan decomposition 100, **103**, 104, 111
    real Jordan form 105, 108, 109

LC system, 39, 45
linear complementary system, *see* LC system
linear matrix inequality, 16
linear program, 10, 15, 16
LMI, *see* linear matrix inequality
lookup table, *see* closed-form solution
LP, *see* linear program
Lyapunov function, *see* sability
Lyapunov stability, *see* sability

mathematical program, 14, 16
max-min-plus-scaling system, 39
MILP, **16**

minimal representation, **8**, 61
mixed logical dynamical system, *see* MLD system
mixed-integer linear program, *see* MILP
MLD system, III, 39, 45
mode, II, 40
mp-LP, III, **18**, 18, **19** 19, 20 50, 53, 46
mp-MILP, III, 46, 47
mp-QP, 18, 19, 46
MPC, *see* model predictive control
MPT, 9, 10, 41, 47
Multi-Parametric Toolbox, *see* MPT

neighborhood, *see* set | neighborhood
norm
    matrix ∞-norm, XVII, **13**, **101**
    matrix 1-norm, XVII, **13**, **89**
    maximum column-sum norm, *see* matrix 1-norm
    maximum row-sum norm, *see* matrix ∞-norm
    vector ∞-norm, XVI, **13**, 34, 101
    vector 1-norm, **13**, 34

performance index, *see* cost | cost function
point-location problem, IV, 30, 31
polyhedron, **8**, 8, 9, 19, 53, 57, 58
    edge, **8**
    face, XV, **8**, 8, 11
    facet, **8**, 11
    normalized, 8
    polyhedral partition, **10**, 12, 57, 59, 61, 84, 94, 120, 128, 143, 157, 160–162
    ridge, **8**
    vertex, 8, 11
polytope, XV, **8**, 9–12, 100, 105, 107, 110
    H-polytope, 9
    V-polytope, 9
    reduction, 8
    P-collection, *see* polytope collection
    polyhedral partition, *see* polyhedron
    polytope collection, XV, 7, 9, **10**, 11
    polytope family, 10
power diagrams, 161
prediction horizon, XVIII, 28, 32, **33**, 36–38,
projection, XVI, **11**
    affine projection, 11
    orthogonal projection, 11
PWA system, I, 31, **39**, 40, 45, 46

Index

QP, *see* quadratic program
quadratic program, **16**, 16, 17

receding horizon control, III, IV, **26**,
    26–32, 34–38, 46, 61, 62, 66
    76, 99, 101, 105, 118–121, 127,
    133, 137, 143, 157, 160, 167
region of attraction, *see* set | region of
    attraction
RHC, *see* receding horizon control

SDP, *see* semidefinite program
semidefinite program, **16**
sequential search, 157,161–164, 169, 170
set
    affine dimension, XV, 5, 8
    affine set, **4**
    ε-ball, XIV, **4**, 4, 5, 9
    boundary, XV, **6**, 11, 38
    bounded set, **6**, 8, 9
    closed set, **6**, 6, 8, 11, 19
    closed set, half-, 6
    closure, XV, 6, 10, 11, 40, 52, 56,
        107, 120, 142, 143, 159
    collection, 7
        polytope, *see* polytope
    compact set, **6**, 40, 51, 52, 68, 72,
        127, 158, 159
    convex set, **3**, 7, 8, 13, 57, 58, 86, 107
    difference, 11
    dimension, XV, **4**, 5, 8
    extended real vector space, 6
    family, *see* set | collection
    full-dimensional, 5
    input admissible set, XVIII, 22, 37
    interior, XV, **5**, 5, 11, 38
        relative, XV, 5
    invariant set, control (maximal),
        22
    invariant set, positively (minimal/
        maximal), IV XVIII, **21**, 22, 24,
        37, 62, 114, 118, 121–123, 125,
        127, 128, 131–134, 137, 138
    Lyapunov stability region, *see*
        stability
    neighborhood, XIV, **4**, 6, 24, 36, 71,
        84–86, **115**, **144**
    open set, 4, **5**, 5, 6 11, 76
    open set, half-, 6
    partition, 7
        polyhedral, *see* polyhedra | polyhedral
            partition
    region of attraction
        maximal, **22**
    region of attraction, XVIII, **22**, 122,
        123, 128, 129, 131, 133, 137
    terminal target set, XVIII, 33, 51,
        62, 114, 115
    underlying set, XV, **7**, 11
set membership problem, *see*
    point location problem
stability, 35, 41, 46, 99, 103,
    asymptotic stability, IV, 22, **23**,
        23–25, 36, 37, 38, 69–71, 73, 82,
        100, 114, 123, 127, 128, 131
    control Lyapunov function, **24**, 24,
        38, 62, 66, 79
    exponential stability, IV, **23**, 23, 24,
        36, 37 70, 73, 115, 145
    Lyapunov function, III, IV, **23**, 23,
        25, 73, 82, 83, 90, 91, 99–102,
        105–109, 11–117, 123, 127,
        128, 134, 143
    Lyapunov stability, **23**, 71, 101, 128
    Lyapunov stability region, 119,
        122, 123, 127–129, 131, 134
state feedback control law, 52
stationarity, 75–78, 122, 132, 133
    stability tube, IV, XIX
steer-to-the-origin problem, 41

value function, 68
vertex enumeration, 12

Printing: Mercedes-Druck, Berlin
Binding: Stein+Lehmann, Berlin

Lecture Notes in Control and Information Sciences

Edited by M. Thoma, M. Morari

Further volumes of this series can be found on our homepage:
springer.com

Vol. 359: Christophersen F.J.
Optimal Control of Constrained Piecewise Affine Systems
190 p. 2007 [978-3-540-72700-2]

Vol. 358: Findeisen R.; Allgöwer F.; Biegler L.T. (Eds.):
Assessment and Future Directions of Nonlinear Model Predictive Control
642 p. 2007 [978-3-540-72698-2]

Vol. 357: Queinnec I.; Tarbouriech S.; Garcia G.; Niculescu S.-I. (Eds.):
Biology and Control Theory: Current Challenges
589 p. 2007 [978-3-540-71987-8]

Vol. 356: Karatkevich A.
Dynamic Analysis of Petri Net-Based Discrete Systems
166 p. 2007 [978-3-540-71464-4]

Vol. 355: Zhang H.; Xie L.
Control and Estimation of Systems with Input/Output Delays
213 p. 2007 [978-3-540-71118-6]

Vol. 354: Witczak M.
Modelling and Estimation Strategies for Fault Diagnosis of Non-Linear Systems
215 p. 2007 [978-3-540-71114-8]

Vol. 353: Bonivento C.; Isidori A.; Marconi L.; Rossi C. (Eds.)
Advances in Control Theory and Applications
305 p. 2007 [978-3-540-70700-4]

Vol. 352: Chiasson, J.; Loiseau, J.J. (Eds.)
Applications of Time Delay Systems
358 p. 2007 [978-3-540-49555-0]

Vol. 351: Lin, C.; Wang, Q.-G.; Lee, T.H., He, Y.
LMI Approach to Analysis and Control of Takagi-Sugeno Fuzzy Systems with Time Delay
204 p. 2007 [978-3-540-49552-9]

Vol. 350: Bandyopadhyay, B.; Manjunath, T.C.; Umapathy, M.
Modeling, Control and Implementation of Smart Structures 250 p. 2007 [978-3-540-48393-9]

Vol. 349: Rogers, E.T.A.; Galkowski, K.; Owens, D.H.
Control Systems Theory and Applications for Linear Repetitive Processes 482 p. 2007 [978-3-540-42663-9]

Vol. 347: Assawinchaichote, W.; Nguang, K.S.; Shi P.
Fuzzy Control and Filter Design for Uncertain Fuzzy Systems
188 p. 2006 [978-3-540-37011-6]

Vol. 346: Tarbouriech, S.; Garcia, G.; Glattfelder, A.H. (Eds.)
Advanced Strategies in Control Systems with Input and Output Constraints
480 p. 2006 [978-3-540-37009-3]

Vol. 345: Huang, D.-S.; Li, K.; Irwin, G.W. (Eds.)
Intelligent Computing in Signal Processing and Pattern Recognition
1179 p. 2006 [978-3-540-37257-8]

Vol. 344: Huang, D.-S.; Li, K.; Irwin, G.W. (Eds.)
Intelligent Control and Automation
1121 p. 2006 [978-3-540-37255-4]

Vol. 341: Commault, C.; Marchand, N. (Eds.)
Positive Systems
448 p. 2006 [978-3-540-34771-2]

Vol. 340: Diehl, M.; Mombaur, K. (Eds.)
Fast Motions in Biomechanics and Robotics
500 p. 2006 [978-3-540-36118-3]

Vol. 339: Alamir, M.
Stabilization of Nonlinear Systems Using Receding-horizon Control Schemes
325 p. 2006 [978-1-84628-470-0]

Vol. 338: Tokarzewski, J.
Finite Zeros in Discrete Time Control Systems
325 p. 2006 [978-3-540-33464-4]

Vol. 337: Blom, H.; Lygeros, J. (Eds.)
Stochastic Hybrid Systems
395 p. 2006 [978-3-540-33466-8]

Vol. 336: Pettersen, K.Y.; Gravdahl, J.T.; Nijmeijer, H. (Eds.)
Group Coordination and Cooperative Control
310 p. 2006 [978-3-540-33468-2]

Vol. 335: Kozłowski, K. (Ed.)
Robot Motion and Control
424 p. 2006 [978-1-84628-404-5]

Vol. 334: Edwards, C.; Fossas Colet, E.; Fridman, L. (Eds.)
Advances in Variable Structure and Sliding Mode Control
504 p. 2006 [978-3-540-32800-1]

Vol. 333: Banavar, R.N.; Sankaranarayanan, V.
Switched Finite Time Control of a Class of
Underactuated Systems
99 p. 2006 [978-3-540-32799-8]

Vol. 332: Xu, S.; Lam, J.
Robust Control and Filtering of Singular Systems
234 p. 2006 [978-3-540-32797-4]

Vol. 331: Antsaklis, P.J.; Tabuada, P. (Eds.)
Networked Embedded Sensing and Control
367 p. 2006 [978-3-540-32794-3]

Vol. 330: Koumoutsakos, P.; Mezic, I. (Eds.)
Control of Fluid Flow
200 p. 2006 [978-3-540-25140-8]

Vol. 329: Francis, B.A.; Smith, M.C.; Willems, J.C. (Eds.)
Control of Uncertain Systems: Modelling, Approximation, and Design
429 p. 2006 [978-3-540-31754-8]

Vol. 328: Loría, A.; Lamnabhi-Lagarrigue, F.; Panteley, E. (Eds.)
Advanced Topics in Control Systems Theory
305 p. 2006 [978-1-84628-313-0]

Vol. 327: Fournier, J.-D.; Grimm, J.; Leblond, J.; Partington, J.R. (Eds.)
Harmonic Analysis and Rational Approximation
301 p. 2006 [978-3-540-30922-2]

Vol. 326: Wang, H.-S.; Yung, C.-F.; Chang, F.-R.
H_∞ Control for Nonlinear Descriptor Systems
164 p. 2006 [978-1-84628-289-8]

Vol. 325: Amato, F.
Robust Control of Linear Systems Subject to Uncertain
Time-Varying Parameters
180 p. 2006 [978-3-540-23950-5]

Vol. 324: Christofides, P.; El-Farra, N.
Control of Nonlinear and Hybrid Process Systems
446 p. 2005 [978-3-540-28456-7]

Vol. 323: Bandyopadhyay, B.; Janardhanan, S.
Discrete-time Sliding Mode Control
147 p. 2005 [978-3-540-28140-5]

Vol. 322: Meurer, T.; Graichen, K.; Gilles, E.D. (Eds.)
Control and Observer Design for Nonlinear Finite and Infinite Dimensional Systems
422 p. 2005 [978-3-540-27938-9]

Vol. 321: Dayawansa, W.P.; Lindquist, A.; Zhou, Y. (Eds.)
New Directions and Applications in Control Theory
400 p. 2005 [978-3-540-23953-6]

Vol. 320: Steffen, T.
Control Reconfiguration of Dynamical Systems
290 p. 2005 [978-3-540-25730-1]

Vol. 319: Hofbaur, M.W.
Hybrid Estimation of Complex Systems
148 p. 2005 [978-3-540-25727-1]

Vol. 318: Gershon, E.; Shaked, U.; Yaesh, I.
H_∞ Control and Estimation of State-multiplicative Linear Systems
256 p. 2005 [978-1-85233-997-5]

Vol. 317: Ma, C.; Wonham, M.
Nonblocking Supervisory Control of State Tree Structures
208 p. 2005 [978-3-540-25069-2]

Vol. 316: Patel, R.V.; Shadpey, F.
Control of Redundant Robot Manipulators
224 p. 2005 [978-3-540-25071-5]

Vol. 315: Herbordt, W.
Sound Capture for Human/Machine Interfaces: Practical Aspects of Microphone Array Signal Processing
286 p. 2005 [978-3-540-23954-3]

Vol. 314: Gil', M.I.
Explicit Stability Conditions for Continuous Systems
193 p. 2005 [978-3-540-23984-0]

Vol. 313: Li, Z.; Soh, Y.; Wen, C.
Switched and Impulsive Systems
277 p. 2005 [978-3-540-23952-9]

Vol. 312: Henrion, D.; Garulli, A. (Eds.)
Positive Polynomials in Control
313 p. 2005 [978-3-540-23948-2]

Vol. 311: Lamnabhi-Lagarrigue, F.; Loría, A.; Panteley, E. (Eds.)
Advanced Topics in Control Systems Theory
294 p. 2005 [978-1-85233-923-4]

Vol. 310: Janczak, A.
Identification of Nonlinear Systems Using Neural Networks and Polynomial Models
197 p. 2005 [978-3-540-23185-1]

Vol. 309: Kumar, V.; Leonard, N.; Morse, A.S. (Eds.)
Cooperative Control
301 p. 2005 [978-3-540-22861-5]

Vol. 308: Tarbouriech, S.; Abdallah, C.T.; Chiasson, J. (Eds.)
Advances in Communication Control Networks
358 p. 2005 [978-3-540-22819-6]

Vol. 307: Kwon, S.J.; Chung, W.K.
Perturbation Compensator based Robust Tracking Control and State Estimation of Mechanical Systems
158 p. 2004 [978-3-540-22077-0]